TOOLING METHODS AND IDEAS

TOOLING METHODS
and IDEAS

Edited by HAROLD W. BREDIN

INDUSTRIAL PRESS INC., 200 Madison Avenue, New York 10016

TOOLING METHODS AND IDEAS

Library of Congress Catalog Card Number: 67-17406

Preface

Tooling is often more an art than a science. In the past, the secret techniques of the tooling "artist" gave a trade advantage to many a successful business. Even the man in the shop jealously guarded his little secrets of the trade from fellow mechanics in the hope of economic gain. Apprentices, on completion of their time, were wisely urged to move from shop to shop to improve their skill and tooling know-how. In this unfortunate atmosphere, the art often died with the artist.

Today, the climate is changing. Technical books, trade journals and professional and trade associations have done much to promote an interchange of information. Still, there remains a notable sluggishness in the spread of tooling ideas. Techniques that make an operation simple in one shop are too often unknown in another. There are few fields where a cross fertilization of ideas is more sorely needed and, ironically, more potentially profitable.

This book is literally a "treasury of practical ideas" for the machinist, the tool and die maker, the draftsman, the tool designer and the production engineer. As a source book of workable tooling methods, it can prove equally useful to the apprentice, the technical and trade school student, the journeyman mechanic or the professional engineer. These are the best of MACHINERY'S Tool Engineering Ideas, arranged in a shop-oriented format designed to simplify intelligent "mining" of the information presented.

The ideas are grouped according to their actual use, but it must be remembered that they may also suit an entirely different operation with similar basic tooling requirements. Obviously, work-holding techniques applied to milling may logically suggest application to lathe faceplate operations or drill jigs. Other tooling adaptations will undoubtedly be more subtle, yet they are literally limited only by the ingenuity of the reader. In today's economy, one idea successfully transplanted from this book can save many times its cost. So good hunting!

Old Tappan, New Jersey   HAROLD W. BREDIN
January, 1967

v

Contents

CHAPTER I

Special Tooling for Milling and Shaping

Spherical Surfaces Milled with Fly Cutters

In the manufacture of parts having the composite form of a cylinder, a partially relieved cone, and a sphere, economical machining of the spherical surface presented a problem. The use of fly cutters and a turntable, however, permits the production of the entire spherical portion of the workpiece in a single operation on a vertical milling machine. The setup is illustrated in Fig. 1.

The part A is mounted on a wedge-shaped block B secured to a power driven turntable C. This block holds the workpiece so that its axis is at an angle X with the vertical, and the center of the spherical portion is positioned on the rotational axis of the turntable. The machine spindle

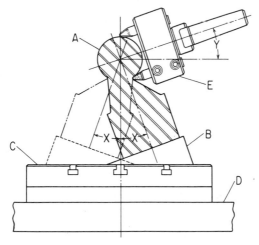

Fig. 1. Setup for machining the spherical portion of a workpiece. The part A is rotated on a power operated turntable C as the fly cutters mill the sphere.

is inclined at an angle *Y* and the milling machine table *D* is positioned so that the axis of the spindle will pass through the center of the sphere at the completion of the operation. A fly cutter bar *E*, with inserted cutters set to cut the proper diameter, is employed. Angles *X* and *Y* and the setting of the cutters are determined by the size and shape of the part.

For the operation, the turntable is slowly rotated under power and the work is fed horizontally toward the cutter. If the machine has a spindle feed, the cutters may be fed axially to the work. A tool setting gage will simplify readjustment of the fly cutters after grinding.

Milling Depth Regulator Which Permits Different Work Sizes

In order to produce in quantities conical parts that varied in size but with finished configurations that were in ratio to each other, it was necessary to prepare for a special milling operation. This was accomplished by providing a standard pantograph type milling machine with a holding fixture and power-driven master. To automate the machine for this application required a special milling depth regulator mechanism with a simple yet precise adjustment to facilitate setups between part sizes. This mechanism is shown in Figs. 2 and 3. The beginning and ending of a milling cycle is shown in Fig. 2.

The mechanism includes a rack *A* which is reciprocated by air cylinder *B*. Stroke length *X* of rack *A* is limited by adjustable stop *C*. The movement of rack *A* toward the right causes rotation of pinion *D*, which is mounted on shaft *E*. Shaft *E* runs in fixed members of the mechanism housing. Rotation of pinion *D* causes shaft *E* to revolve.

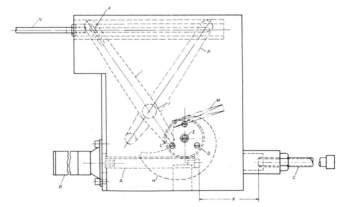

Fig. 2.   Mechanism designed to permit changes in cutter depth on a pantograph type milling machine.

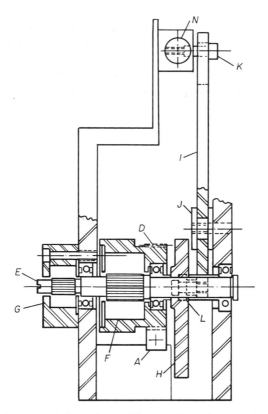

Fig. 3.   End view of the milling depth regulator in Fig. 2.

The shaft is permitted to rotate in the counterclockwise direction only by the holding force of clutch F (Fig. 3), and the slipping of clutch G. Rack A is retracted after each forward stroke in readiness for the next actuation. When the rack is retracted clutch G holds shaft E stationary, while clutch F slips about shaft E. These one-direction clutches F and G are arranged to permit manual operation of shaft E by the application of a screwdriver to the slot in the left-hand end. Thus, in the event of cutter trouble, the operator may change the cutter and the setup at the beginning of a cutting cycle.

Cam H is keyed to shaft E, and so the cam is rotated with each stroke of rack A. The face of the cam applies its force on roller L, which is fastened to lever I. The latter swivels about fixed fulcrum J. Spring M holds roller L against the face of the cam. When the lever is moved to the right by the action of the cam, rod N is operated to the left on a fixed horizontal plane through pin K. Rod N is fixed to the milling head (not shown) and thus controls the milling depth of each pass of the cutter. Each pass begins with a stroke of rack A. Position P shows the maximum stroke of both cam H and lever I.

Milling depth is set up by using gage-blocks to set stroke length X. By merely increasing or decreasing the stroke which controls the cutting

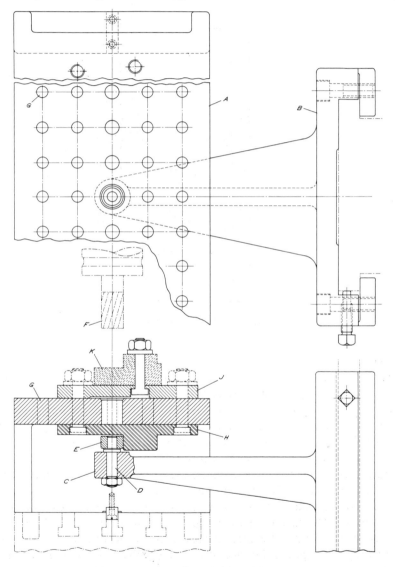

Fig. 4.   Holes G in box table A permit template H and work fixture J to be tied together in the proper relationship.

depth of the milling cutter, all part sizes can be automated without any other change-over.

Roll Follower for a Vertical Milling Machine

A roll follower device that features an auxiliary box table simplifies working from a template on a column and knee type vertical milling machine. As can be seen in Fig. 4, the box table A has an open side; and, when mounted on the machine table, the open side faces the column.

Clamped to the vertical knee slide is a bracket B, the arm of which projects through the open side of the box table. In constructing the device, a boss C on the end of the arm is bored from the machine spindle to receive a roll stud D. A roll E, having an outside diameter corresponding to that of the end mill F, is fitted over the top of the stud. A series of equally spaced jig-bored holes G permits matched templates beneath the table and work fixtures above the table, such as H and J, respectively, to be tied together by dowel-pins and bolts in proper relationship. The workpiece K, in turn, is secured in place by a bolt passing upward through a T-slot in the fixture.

In use, the operator adjusts the machine table until the template contacts the roll, then employs coordinate movements of the roll-feed and cross-feed handwheels to follow around the roll. Although the template is not visible to the operator, he soon acquires the proper "feel." For roughing cuts, a roll is used that has a slightly larger outside diameter than the end mill.

Multiple-Spindle Milling Head

A multiple-spindle head applied to the spindle of a horizontal boring, drilling, and milling machine permits the simultaneous operation of three end-mills. The device, shown in Fig. 5, was developed for a press manufacturer to machine T-slots in bolster plates too large to be handled on the milling machines available in the plant.

A mounting plate A is doweled and screwed to the end of the machine spindle. The welded housing B is permanently fastened to an adapter C. Two locating pins D in the adapter engage corresponding holes in the plate, aligning the head properly when mounted.

Within the housing, three auxiliary spindles E, F and G are arranged in tandem. Each spindle is borne by a tapered roller bearing H at the driving end and a ball bearing J near its rear end. The drive from the machine spindle K is transmitted through a floating coupling L to the

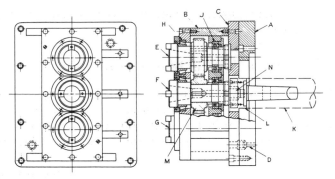

Fig. 5. This multiple-spindle head permits the simultaneous milling of three T-slots in work set up on a horizontal boring, drilling and milling machine.

central auxiliary spindle *F*. A gear *M* fitted around this spindle meshes with an identical gear around each of the other two auxiliary spindles *E* and *G*. Thus, the central spindle drives the two other spindles in a reverse direction.

All three spindles have American Standard tapers, and accommodate stock cutter-arbors. Draw-bolts *N* secure the arbors in the spindles. In use, spindle *F* carries a left-hand end-mill, and spindles *E* and *G* carry right-hand end-mills, since the machine spindle runs clockwise (as viewed looking toward the spindle).

Production Rack-cutting on Horizontal Milling Machine

Although it is possible to cut racks on a milling machine by the use of a rack-cutting attachment and a single-tooth form cutter, this method is inadequate if more than a few units are required. The setup shown

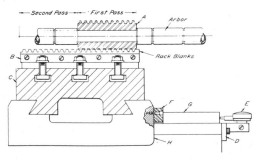

Fig. 6. Special indexing arrangement, together with multiple-tooth annular cutter and simple fixture, permits the production milling of teeth on long racks.

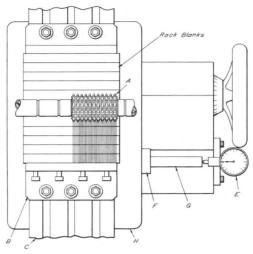

Fig. 7.   A group of 12-inch long racks are cut in two passes with a 6-inch wide cutter.
Spacer G is removed for the second pass.

in Figs. 6 and 7 is designed to facilitate the production milling of racks, 1 foot in length, on a horizontal milling machine.

The only item of tooling that requires any considerable investment is a multiple-tooth annular cutter A, Fig. 6. It is a one-piece unit, 6 inches wide. Because of its width it can cut the 12-inch row of teeth in two passes. A simple plate type fixture B, which is made to hold a number of steel rack blanks, is bolted to the milling machine table C.

The principal difficulty in this type of operation is repositioning the rack for cutting teeth on the second half of the blanks. However, this problem has been handled in the following simple manner.

Bracket D is screwed to the outer end of the machine knee to support dial indicator E. A hardened and ground stop-block F is screwed to the outer face of the machine saddle. The purpose of this stop-block is to assure accurate contact with the indicator point and also with a 6-inch gage-block G.

Prior to making the first pass, the saddle H is moved toward the machine head a sufficient amount to permit gage-block G to be placed between the stop-block and the indicator contact point. The indicator is then set to zero and the cut taken, as shown in Fig. 7. Only one pass is necessary across each half of the rack blanks as the machine is set to cut the full depth of tooth.

When this cut is completed, the saddle is backed off slightly and the gage-block removed. The saddle is then moved outward until the indicator arm contacts stop-block *F*, and the needle is once again at zero. In this position the cutter and workpieces are accurately aligned for the second and final pass which completes the job.

While this setup involved only two passes, there is no reason why longer racks requiring additional passes of the cutter could not be accommodated. The only alteration in the tooling would be to provide additional gage-blocks of the proper lengths. Total length of the racks would, of course, be limited by the extent of the saddle movement.

Instead of using the solid cutter mentioned, it is possible to use a gang of single-tooth form cutters provided with appropriate spacers between them This arrangement would probably be less expensive. It would, on the other hand, be less convenient to handle and certainly not as foolproof.

Vise Supported from Dividing Head for Compound-Angle Milling

A shell end-mill arbor and an adapter block can be used to support a small drill vise on the spindle of a universal dividing head. Such a setup is helpful in milling compound angles on small tool and gage components, particularly of the type illustrated in Fig. 8.

Fig. 8.   A compound angle is milled in a workpiece held in a vise which, in turn, is supported from a dividing head.

Fig. 9.   This special adapter block and standard shell end-mill arbor supports the vise on the spindle of the dividing head.

The arbor and block appear separately in Fig. 9. On its top, the block has a channel which receives the base of the vise. Set-screws in one wall of the channel lock the vise in position. The bottom of the block has a slot which fits the two driving lugs of the arbor. A cap-screw runs through a counterbored hole in the center of the block, securing it to the end of the arbor.

Turn-milling on a Standard Horizontal Milling Machine

Turn-milling is a procedure involving the rotation of a workpiece in contact with a rotating milling cutter, or gang of cutters. This process may be used to advantage in machining interrupted surfaces such as gear segments and other partly circular components. Grooves can be finished in one revolution of the work by plunge-cutting with a cutter of appropriate width. The numerous cutting edges on a milling cutter assure a longer life than could be expected from a single-point turning tool.

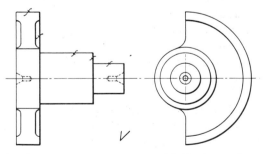

Fig. 10.   Diagram of a forged-steel gear segment, indicating surfaces machined in one operation by turn-milling.

The workpiece V shown in Fig. 10, part of the kick-starting mechanism on a motorcycle, is a steel forging. While turning the parts on a turret lathe, it was found that the interrupted surface of the gear segment caused the cutting edges of the turning tools to deteriorate rapidly so that tool maintenance time was high. Turn-milling was then successfully employed for rough-machining all the diameters and faces indicated in Fig. 10, using the power-driven device shown in Fig. 11. Prior to turn-milling, the forging is centered at both ends.

Since this process requires that the work be revolved, the driving arrangement seen in Fig. 11 is coupled to the horizontal milling machine mandrel. Worm A is secured to the horizontal cutter mandrel B in a manner similar to that used for mounting the cutter spacing collars. The worm meshes with a bronze worm-wheel C which is suspended from the over-arm D by means of a bracket E. This bracket is built up from an anchor block and two side-plates.

Rotation of the worm-wheel C is transmitted to the horizontal splined shaft F, which can slide axially within the bore of the worm-wheel. Integral with the splined shaft is a second worm G. The worm and its splined shaft are supported in a bearing H and retained by means of split nut J, thereby allowing the endwise slack between the shaft and its bearing to be taken up. Rotation from the second worm G is transmitted to the worm-wheel K which is secured to the end of the work-spindle L. It is apparent that the foregoing arrangement is in the form of a double-reduction worm-gear enabling the work-spindle to be rotated by the milling machine cutter mandrel at a slow relative speed. The splined shaft permits the work-spindle to be moved either toward, or away from the cutters.

A flat circular plate M, with an integral head-stock center, is fitted to the work-spindle L. The spindle is housed within a solid-steel head-stock block N. The drive is transmitted from the work-spindle to the workpiece V by means of a pair of block-shaped dogs Y on the circular plate M. These dogs engage the diametral flat on the gear segment. Clearance is provided between the dog faces and the flat on the gear segment to allow for deviations in the forgings. Therefore, only one of the dogs is in driving engagement during the turn-milling operation. The second dog merely prevents the workpiece from swinging on the centers during loading.

A tailstock block P carries a carbide-tipped center. This allows a firm pressure to be applied to the work and eliminates the dangers of scoring and seizing of the tailstock center and the center hole in the parts. The tailstock center has a parallel shank threaded externally which engages a ring nut Q. The center can be advanced into the work

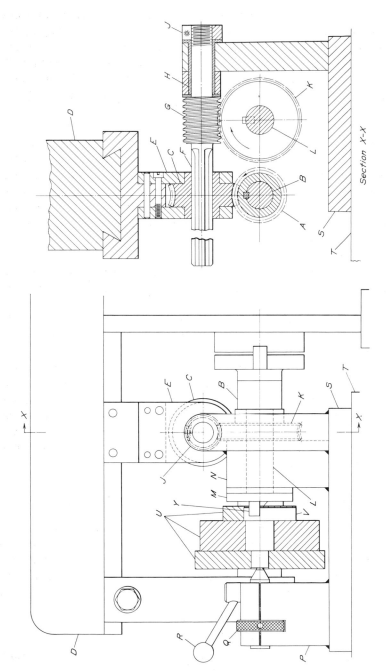

Fig. 11.   Turn-milling attachment incorporating a double worm-gear reduction designed for a standard horizontal milling machine.

by rotating the ring nut. A horizontal split on one side of the tailstock block can be clamped together by means of clamp *R*, thus gripping the tailstock shank firmly. The headstock and tailstock blocks, or housings, are welded to a steel baseplate *S* which is bolted to the machine table *T*. This mounting enables the operator to advance or retract the workpiece in relation to the cutters during the operation by traversing the table.

In use, the table is unlocked and moved longitudinally by means of the quick traverse to bring the work away from the cutters. The tailstock center is unlocked and retracted by means of the ring nut *Q* so that the workpiece can be removed. A new workpiece is then locked between the centers and the table advanced toward the cutters. The cutters *U* are fed into the work *V* by hand until the correct depth has been reached; then the table is locked in position. One revolution of the work in contact with the cutters is sufficient to machine the three diameters and two faces to size.

The mandrel speed is governed by the peripheral speed of the largest-diameter cutter, while the feed, or revolutions per minute of the work-piece, is governed by the chip load per tooth of the smallest-diameter cutter. Therefore, the speed reduction between the mandrel and work-spindle must be controlled closely in the design of the worm gearing to obtain satisfactory cutter life.

Climb-milling has proved very satisfactory for this setup due to the small amount of backlash in the worm-gear drive, thus avoiding the tendency for the cutter to grab at the work. In climb-milling, the surface of the work being cut is moving in the same direction as the teeth doing the cutting, thereby subjecting the gear train to far less strain than would be experienced if it were called upon to drive the work against the cutter rotation.

Milling Two Different Radii on One Curved Surface

Milling the arc-shaped outer edge of the brass casting shown at X in Fig. 12 presented numerous difficulties in regard to holding and feeding the work. The edge to be machined consists of two blended curved portions of different radii, the longer section being a 9-inch radius, and the shorter, a 4-inch radius. Along the center of this curved rim is a shallow rectangular groove of a similar contour.

Member *A* of the fixture shown at Y is a right-angle cast-iron base bracket which is machined all over. The horizontal leg of the bracket contains four drilled holes to receive standard hold-down bolts. Locating tongues to engage the table T-slots may be provided.

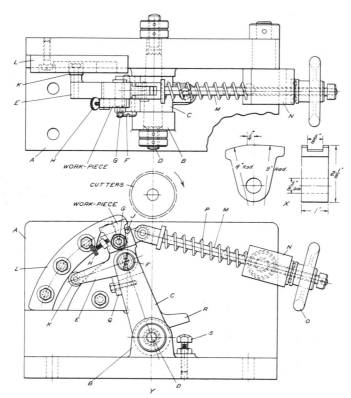

Fig. 12.   Milling fixture designed for machining a curved surface composed of two different radii.

A groove is milled across integral cast boss *B* to receive the lower end of steel arm *C*. This arm pivots on a hardened and ground tapered shaft *D* which fits in tapered holes in both boss cheeks. Mounting the arm on a tapered shaft permits making adjustments to compensate for wear, thus eliminating any play that might develop.

The upper end of the arm is slotted to receive L-shaped lever *E*. The two members are held together by means of a shoulder stud *F*, the reduced threaded end of which screws into a hole in one side cheek of the arm slot. Lever *E* must have a minimum amplitude of movement of 90 degrees independent of the arm movement.

Correct radial positioning of the workpiece is obtained by slipping the previously bored 5/8-inch diameter hole over hardened stud *G*. This stud is mounted on the upper arm of lever *E*. A hexagon nut threaded on the end of the stud serves to lock it against the lever. Radial locking is accomplished by turning knurled-head screw *H* to force the part against stop-pin *J.*

Steel roller *K* rides within a dual-curved track formed in the front face of steel plate *L.* The plate is secured in a recess in the upright base member. The radii of the two curved portions of the roller track correspond to the radii of the curvatures to be milled, but their relative lengths are not critical.

A lug on the right-hand side of lever *E* is pinned to the shackle end of rod *M.* At its opposite end, the rod passes through a clearance hole in a rectangular trunnion block *N* This member has an integral shaft which fits in a bearing hole bored through the upright wall of the base bracket. A cross-pinned collar retains it in place, permitting a swiveling movement only.

Handwheel *O* is threaded on rod *M.* A standard ball thrust bearing is inserted between the handwheel and the trunnion block. To prevent the handwheel from being completely backed off, a mild-steel collar is pinned to the end of the rod. Coil spring *P* is situated over the rod so that it tends to move both the rod and the arm toward the left.

Two screws hold rectangular stop-plate *Q* against the left-hand side of arm *C.* The lower arm of lever *E* contacts the stop-plate when the centers of shaft *D* and studs *F* and *G* are in line.

A lug *R*, integral with arm *C*, is a stop for limiting the amount of clockwise movement imparted to the arm. At a certain point in the cycle of movements, the under side of this lug strikes the head of an adjustable stop-screw *S.* Its proper adjustment must be determined by trial and error.

After the workpiece is securely mounted in the fixture, the handwheel is slowly rotated so as to draw rod *M* and connected members to the right. Under this movement, arm *C* will swivel in a clockwise direction, lever *E* moving in unison with it because of constraint applied by the engagement of roller *K* within the lower portion of the curved track. Accordingly, the workpiece is fed beneath a gang of three cutters, the center one being used to form the shallow groove.

The positions assumed by the various working members of the fixture when the 9-inch radius has been completed may be seen at X in Fig. 13. The radius of the milled portion of the workpiece will be equal to the distance from its rim to the center of shaft *D.*

As rotation of the handwheel continues, arm *C* remains stationary due to the contact of lug *R* and set-screw *S.* At this point, lever *E* swivels about stud *F.* This swiveling action occurs as the roller rides in the shorter portion of the curved track.

The position of the fixture at the completion of the milling operation is shown at Y, Fig. 13. This second phase of the milling cycle produces

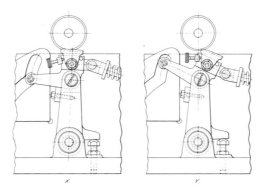

Fig. 13.   Fixture is shown at X after milling a 9-inch radius. The position of fixture
components at the completion of the operation may be seen at Y.

a curvature on the rim of the workpiece of a radius equal to the distance
from the rim to the center of stud *F*. Thus, with a single continuous
movement of the handwheel, both radii are quickly machined.

Concave and Convex Machining on a Shaper

Machining a concave or convex surface can become a troublesome
problem in shops not specifically equipped for the performance of
such operations. Consequently, such operations can become costly in a
job shop.

Figure 14 shows a simple, inexpensive way to produce such surfaces
on an ordinary shaper. Cutting tool *A* is mounted in a hinged coupler *C*
which is attached to bar *B*. This bar is firmly secured to the shaper ram.

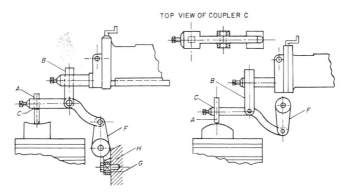

Fig. 14.   Diagrams illustrating methods of machining concave and convex surfaces on
a shaper.

Coupler *C* is guided at two points, being pivoted on the reciprocating bar *B* and guided in an arc by link *F*. The latter rocks on a bearing held in a block fastened to the machine frame. The reciprocating motion of the shaper will cause the cutter *A* to move in a curve under the control of the coupler.

The illustration at the left of Fig. 14 shows the arrangement necessary for cutting a concave surface. For a convex surface, link *F* would be reversed as shown in the diagram at the right. By changing the center-to-center dimensions of parts *A*, *C*, and *F*, a wide variety of curves can be machined.

Arbor Stabilizer for Horizontal Milling Machine

A horizontal milling machine will sometimes set up a bad chatter when a peripheral cutter must be arranged near the midpoint of a long arbor. The consequence is a poor finish of the workpiece, most likely occurring on an old machine having pronounced wear in the spindle bearings — particularly where there is no regular intermediate arbor support on hand. The stabilizer in Fig. 15 employs a roller pressure principle and is intended to correct this unfavorable condition.

Two V-shaped yokes *A*, one on either side of a cutter such as *B*, are welded to a split bushed liner *C* located on the over-arm *D* of the machine. The bottoms of the yokes have a milled channel *E* to receive the threaded studs *F* of cross-bars *G*. Hardened steel rollers *H* bear against the arbor spacing collars *J*, and are pinned to arms *K*. These arms are suspended at the ends of the cross-bars, and work around links *L* extending from the cross-bars. Necessary roller pressure is obtained by adjusting the position of a lock-nut *M* on the threaded stud *F*.

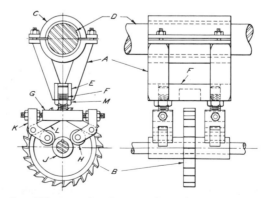

Fig. 15.   The rollers *H* bear on the spacing collars of the arbor to eliminate chatter.

The position of the rollers enables them to absorb both lateral and vertical thrusts. Also, their position above the arbor center line leaves the cutter free to operate at the desired depth without obstruction. Cutters can be changed with a minor disassembly of the stabilizer, it being necessary only to loosen the lock-nut M on the front cross-bar stud, and to slide this unit out of the channel E.

For milling machines having double over-arms, or over-arms that are rectangular in cross-section, the yokes A and liners C will have to be designed accordingly.

CHAPTER 2

Fixtures for Milling and Shaping

Fixture for Milling Curved Slots without Rotary Table

Milling a curved slot in a small quantity of large, sector-shaped steel weldments presented a problem in a shop where a suitable rotary table was unavailable. The illustrated fixture (Fig. 1), designed for a vertical

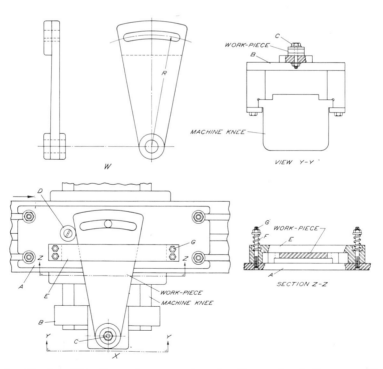

Fig. 1. Simple milling fixture for sector-shaped weldment that facilitates machining of curved slot.

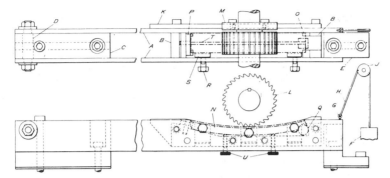

Fig. 2. The counterweight *F* keeps the fixture in a normally raised position, so that the template *K* is in contact with the follower *M*.

milling machine, utilized the motion of the machine table to pivot the weldment about a locating point on a stationary bracket.

At W can be seen the part to be machined. The slot is located from the center of a shaft hole previously bored through the hub. Baseplate *A* of the fixture, which is shown at X, is bolted to the milling machine table. A welded bracket arrangement *B* is machined to permit clamping to the knee of the milling machine (view Y-Y).

Bolt *C*, of the same diameter as the hole in the workpiece hub, passes through the part and serves as a pivot point. The bolt is secured to bracket *B*. Steel roller *D*, which is fastened to the baseplate, contacts the side of the part. Instead of a steel roller, a ball bearing may be used.

To retain the sector against the baseplate, a spring-loaded hold-down bracket *E* is used. This channel-shaped member, which can be seen more clearly in section Z-Z, is held firmly against the surface of the workpiece by four die-springs *F*, supported by four studs *G*.

Before the part is mounted in the fixture, it should be coated with either grease or heavy oil. After mounting, the milling machine table is adjusted to bring the cutter to the desired radial distance from the pivot point, and at the right-hand extremity of the slot to be machined.

When the table feed is engaged in the direction indicated by the arrow, roller *D* will bear against the workpiece and cause it to swing around bolt *C*. The nuts on studs *G* can be adjusted to alter spring tension on bracket *E*, thus providing a means of eliminating chatter.

Pivoting Fixture Used for Milling a Curved Surface

Figure 2 shows a work-fixture which solved the problem of milling a curved surface at a time when no tracer-controlled equipment was available. A horizontal machine set up with an arbor type cutter was used for the job.

In principle, a counterweight is employed to change the position of the work on the milling table during the cut. A template of the required curvature — in this instance, a concave surface — is incorporated in the fixture and maintains contact with a follower that rotates with the cutter.

The fixture consists essentially of front and back plates *A* joined together by three spacers *B*. At their left end, the plates are hinged to an anchor *C* by a bolt pin *D*. An alignment block *E* bolted to the table at the other end of the fixture fits the channel between the plates, but is not tied to them. In this way, a counterweight *F*, connected to an eyebolt *G* by a cable *H* running over a pulley *J* keeps the fixture in a normally raised position.

Attached to the back plate is a template *K* having an upper edge of the required curvature. On the arbor behind the milling cutter *L* and directly over the template is a follower *M*. The follower has the same diameter as the cutter. Actually, the follower is a ball bearing, with its inner race free to rotate at spindle speed while the outer race rolls on the template.

The work *N*, a bracket forging, is nested in the channel between the plates on shoulders *O* of the right-hand spacer and on a bridge piece *P* of the left-hand spacer. The bridge piece is able to swivel to adjust itself to any irregularity of the bottom surface of the forging. After being registered endwise against a locator *Q*, the work is secured by tightening set-screws *R* which operate clamps *S*. A central rib *T* along the convex bottom surface is supported over a pair of screw-jacks *U*.

The cut is taken with a feed of the table to the left. During the cut, the counterweight keeps the template in contact with the follower, causing the plates to pivot slightly on the anchor at their left-hand end.

Hand-operated Radius Milling Fixture

A pin with a rectangular head on which unusual radius milling operations must be performed is shown in Fig. 3. The cylindrical shank *A* of this pin is turned to a diameter of 5/8 inch within close limits, and the rectangular head *B* at one end of the shank is surface-ground to close tolerances on all sides except the one joined to the shank. The two slots *C*, identical in width and depth, are machined in the top of the head portion by straddle-milling cutters. The center axis of each slot is exactly in line with the main axis of the shank. Both slots are located at

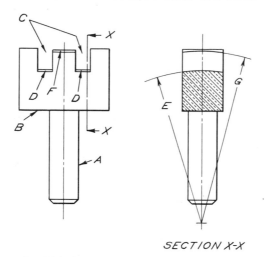

SECTION X-X

Fig. 3.　Component in which slots C are milled by use of fixture shown in Figs. 4 and 5.

right angles to the long sides of the head. The part is made of free-cutting brass.

All dimensions of the two slots must be held to limits of plus or minus 0.002 inch. The bottoms of slots *D* are accurately milled to a radius *E* of 3 inches. The 1/4-inch wide crest *F* is machined to a radius *G* of 3 1/4 inches from the same center.

The quantity of these components to be made, although large enough to necessitate a special holding fixture, did not warrant the production of an elaborate device. The diagrams, Figs. 4 and 5, show the simple but effective milling fixture developed. Although this fixture is designed for hand operation, the short length of the cut taken, and the rapidity with which the parts can be loaded, machined, and unloaded, make it possible to obtain an economical production speed.

The under side of the fixture body *H* is machined for mounting on the milling machine table. A swinging arm member *I* is slotted at its lower end for a close fit over boss *J*. A minimum of side clearance is allowed between these members. The arm is pivoted on the hardened headed steel stud *K*. Thus mounted, the arm should have a certain amount of swiveling movement on each side of the vertical axis of the fixture.

Stud *L* is screwed tightly into the right-hand side of the arm and fitted over it is a slotted steel shackle formed integrally on the end of the cylindrical rod *M*. The shackle and stud are connected by a pivot-pin.

At the right-hand end of body *H* is an upright column with a rectangular slot machined completely through it in line with the axis of stud *L*

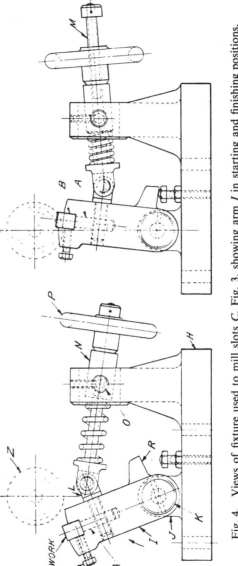

Fig. 4. Views of fixture used to mill slots *C*, Fig. 3, showing arm *I* in starting and finishing positions.

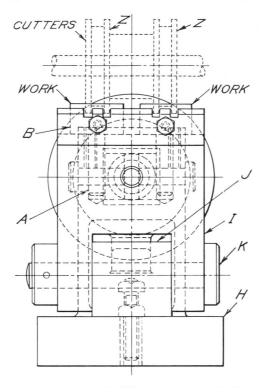

Fig. 5. End view of milling fixture shown in Fig. 3.

when arm I is in the vertical position. Inserted in this slot is a hardened steel rectangular block N, which is made a close sliding fit between the front and rear vertical sides of the slot. An ample amount of clearance should be provided between the top and bottom sides of the slot and the block in order to permit the latter member to swivel.

Block N is pivoted on a pair of headed pins O which are located one at each side of the column on the same centers in both vertical and horizontal planes. The pins are retained in the column by headless screws. Block N is bored lengthwise to be a sliding fit over the rod M, and the pivot-pins O project into the sides of the block only a sufficient amount to prevent interference with the sliding movements of the rod.

Rod M is threaded to take aluminum handwheel P, the boss of which bears against the end of block N. A compression spring encircles the rod between block N and the shackle at the end of rod M, to hold the handwheel tightly against the block, and force arm I to turn in a counterclockwise direction when the handwheel is released. To restrict the adjusting movements of the handwheel on the threaded end of rod M, a small collar is fastened to the right-hand end of the rod.

A rectangular slot extends across the upper end of arm I to receive head B of the work. The center of this slot coincides with the center axis of the arm passing through the pivotal point on stud K. The width of the slot is made about 0.002 inch greater than that of the rectangular head of the component to be milled. The depth of this slot is less than the height of the workpiece head to allow clearance for the milling cutters used to form the slots. Two reamed holes 5/16 inch in diameter in the base of the slot receive the cylindrical shank of the work.

The fixture is designed to take two components at each setting, these being mounted side by side in the slot in arm I, as shown in Fig. 5. Engagement of the shank portions within the holes in the bottom of the slot insures accurate location of the parts in crosswise relationship to the cutters, while the sides of the slot will insure that the rectangular head portions are positioned at right angles to the cutters.

The base of the slot must be exactly parallel with the axis of fulcrum stud K and the machined base of body H when the arm is mounted in the working position, as shown. Each component is retained in the correct position in the arm by a set-screw.

To control the amount of swiveling movement imparted to arm I, this member is provided with a lug R on the right-hand side. This lug bears on the rounded head of a hardened stop-screw in a tapped hole in body H.

After the fixture is fastened to the milling machine table, the table and cross-slides are adjusted so that the center of fulcrum pin K lies exactly on the vertical axis passing through the milling cutters on the arbor, and with the cutters in the correct transverse position relative to the length of the heads. All slides and the table of the machine are then locked. The slots are machined by swinging arm I.

When the slots have been machined halfway through, arm I will have reached the middle vertical position perpendicular with the base. Slots C (Fig. 3) are machined by the pair of cutters Z, (Fig. 5) mounted on the arbor of the milling machine. These cutters are spaced 1/4-inch apart. The cutters form the narrow 1/8-inch wide slots, and also produce the arc at the base of each slot. A third cutter, of smaller diameter, is mounted between cutters Z to produce the curved crest F of the head.

The cutters revolve in a clockwise direction and operation of the fixture commences from the position shown at the left, (Fig. 4). In this starting and loading position, arm I has been swiveled counter-clockwise to bring the workpieces clear of the cutters. This position will be determined by the stop-collar at the right-hand end of rod M which makes contact with the boss on the handwheel. As the handwheel is rotated in a clockwise direction, rod M will be drawn gradually through the block N, and at the same time arm I will be slowly swiveled toward the right about pin K. This hand-operated movement is continued until the head B of the pin has been completely traversed by the cutters, and the final stop position is reached when lug R bears against the stop-screw in base H, as seen in the view at the right of Fig. 4.

The threads on rod M should be of fine pitch to permit a suitable rate of feed of the work past the cutters. As arm I undergoes the swiveling movement, the shackle connection allows rod M to adjust itself radially to suit the angularity of the arm at all points throughout its working range. The base of the slot in the end of arm I, which receives head B of the work, should be located the exact distance away from the center of the fulcrum stud so that the necessary radius will be obtained on the bases of the two slots.

Circular Knife-Edges Locate and Grip Thin Castings

Holding a thin casting in a fixture while surface milling can often be a problem, as it usually is not possible to apply clamps against its top surface. One solution to this problem is presented in Figs. 6 and 7.

Fig. 6.   Fixture uses three circular knife-edges B to locate and grip thin casting A while surface milling.

Fig. 7. Sharp-pointed screws *C*, positioned at an angle, force the workpiece against the side locators and down on the fixture base.

A small cover-plate casting *A*, Fig. 6, is positioned against three circular knife-edge locators *B*. Two hardened screws *C*, their ends ground to sharp points, pass through block *D* at a downward-pointing angle. The sharp points contact the casting, forcing it against the two side locators while seating it firmly against the fixture base. The third knife-edge is at the end of the casting to oppose the feeding force of the cutter.

Knife-edge locators of this type will bite into the surface of the casting and prevent it from lifting during the cut, regardless of the draft angle encountered. They should be made of tool steel, then hardened and tempered to approximately 58 to 60 Rockwell C.

The circular shape is particularly advantageous as it permits periodic rotation of the locators when one portion of the edge becomes dull from repeated use. They are secured to the fixture base by socket-head cap-screws to permit quick release when it becomes necessary to turn them. The fixture is shown removed from the milling machine in Fig. 7.

Indexing Fixture for Milling Thrust Type Washers

A bronze thrust washer, used extensively in the production of transmissions for farm tractors, is shown at *X* in Fig. 8. Because of the thin cross-section of the washer, clamping is difficult if springing is to be avoided and ample clearance for the cutter is to be provided when milling the oil-grooves. The indexing fixture seen at *Y* surmounts these obstacles.

Base *A* of the fixture is slotted at both ends for clamping to the machine table. Beneath the base, and in line with the slots, are two keys *B*, their purpose being to align the base with the table. To the top surface of the base is welded a vertical cylindrical member *C*. This upright member is ground on its upper face and reamed to a slip fit with adapter *D*.

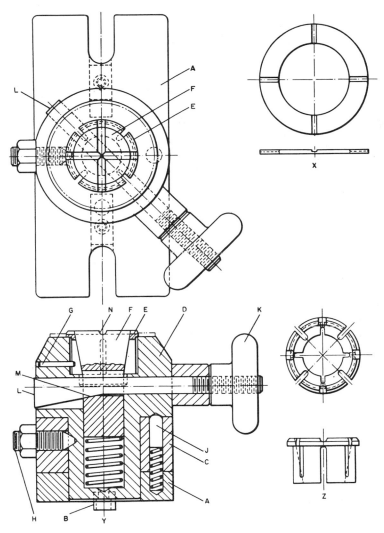

Fig. 8.   Indexing type milling fixture firmly holds the thrust washer shown at X without distorting it. Loading and unloading are rapid.

The adapter is counterbored and reamed to receive collet jaws E and spring-loaded draw-pin F. A detailed drawing of the collet jaws may be seen at Z. Dowel-pin G prevents the jaws from turning.

A cone-point set-screw H is threaded through the wall of cylinder C. The point of the screw extends into a groove that is machined around the adapter shank for a distance equal to a 90-degree arc. A lock-nut retains the screw in position. The rounded end of spring-loaded index-pin J, which is housed within the upright cylindrical member, clicks into either of two notches located in the under side of the adapter flange. These notches are located 90 degrees apart and serve as the indexing points for the adapter assembly.

In use, the fixture is first bolted to the machine table, following which a thrust washer is slipped over the collet jaws. Approximately one-half turn of knob K will pull draw-stud L a distance sufficient to force draw-pin F in a downward direction against a spring. This is accomplished through the action of the mating tapers at M.

The first pass may then be made, using grooves N as a guide for the cutter. When this initial cut is completed, the adapter is indexed 90 degrees, or until index-pin J clicks into the second notch in the adapter flange. A second pass may now be made to complete the workpiece.

Unloading the fixture is extremely simple. By backing off knob K, draw-stud L will move to the left, relieving the pressure on draw-pin F. When this happens, the spring beneath the draw-pin will force it up, thereby allowing the collet jaws to spring together to their original position. The thrust washer may then be lifted from the fixture.

Collet Type Fixture Indexed by Means of Template

A quick-acting milling fixture that incorporates a replaceable template for indexing purposes, and a standard split collet for gripping the workpiece, is shown in Fig. 9. This fixture was designed to simplify setting up, indexing, and milling an accurate seven-sided polygon on the small steel shoulder-stud shown at A. Because it is designed for using interchangeable collets, the fixture is widely applicable.

The cast-iron body B is bored longitudinally to receive a ground sleeve C. A tolerance of plus or minus 0.0005 inch is held on the dimension between the axis of the bore and the machined mounting face on the under side of the body casting.

Fitting the sleeve is a standard split collet D. The left-hand end of sleeve C is bored to match the side taper on the conical head of the collet. A pin, pressed into the ground sleeve, projects into the bore sufficiently

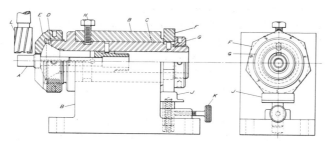

Fig. 9.   Milling fixture that accepts interchangeable collets is indexed by means of replaceable templates.

to engage a keyway provided in the shank of the collet, thus preventing rotation between the two members.

Casehardened steel locking-cap E is threaded internally to engage threads on the extreme left-hand end of the ground sleeve. This cap is knurled on its outside diameter and provided with four equally spaced holes for a spanner wrench. The inside end-wall of the cap is domed to the same curvature as the collet head against which it bears. A clearance hole is drilled through this same end-wall so that the longest shank of the workpiece may be passed through it into the jaws of the collet.

The right-hand end of sleeve C projects from the body casting, and is machined to provide a bearing surface for template F. This template is a push fit on its bearing surface and is restrained from turning by means of a pin and keyway.

In this case, the perimeter of the template is accurately formed with seven equal sides, as shown in the side view at the right in Fig. 9. The template is made six times larger than the corresponding polygonal shape required on the workpiece. This enlargement not only simplifies manufacture but also promotes greater accuracy in the finished product. The completed template should be casehardened and, preferably, lightly ground on all sides to insure flatness. Circular lock-nut G, having spanner holes in its rim for tightening, holds the template in place.

Sleeve C is locked within the body at each of the required seven radial settings by means of set-screw H. The tip of this screw is hardened, and extends into a shallow annular groove cut around the periphery of the sleeve. By allowing the screw to bear down on the base of this groove, resulting burrs, or flats, which are likely to arise from repeated gripping, will have no adverse effect upon the continued smooth working of this sleeve.

The sleeve, and all members mounted thereon, are indexed within the body to the desired seven positions by means of the T-shaped slide J.

This slide has a rectangular head which is appreciably longer than the various sides of the polygonal template. The cylindrical shank of the slide is machined to a medium fit within its mating hole in the body. A compression spring, which is located within the hollow vertical leg of the slide, forces it upward and in contact with the flats on the template.

The surfaces of the rectangular head of the slide are carefully ground for flatness and squareness. It may be observed that the long, rear side of its head is in sliding contact with the vertical side of the body. Fitting the slide in this way prevents the member from turning when lock-screw K is released.

In use, the fixture is bolted to the table of a vertical milling machine. Template F is set so that one of its flats is flush against the top surface of spring-loaded slide J, following which lock-screw K is secured. Set-screw H is then tightened securely so that the sleeve is located in the position.

The workpiece A is inserted through the clearance hole in the locking-cap, into the jaws of collet. To grip the workpiece, the locking-cap is rotated clockwise. This action forces the collet farther into the sleeve, thus causing the jaws to close on the work. A small end-mill L is used for the machining operation.

Milling Fixture Automatically Clamps and Releases Small Work

Automatic clamping and releasing of small workpieces are features of the fixture shown in Fig. 10. This fixture was designed to facilitate milling a burr from one end of cylindrical workpieces left after a lathe operation and holds twenty pieces at a time. In the enlarged view of a workpiece (lower right) an arrow points to the centrally located burr.

The base of the fixture is fitted with standard milling machine keys and V-bolts for aligning on the machine table. Worm A driven by a pulley is geared to worm-wheel B, the latter being bored to fit over turntable E.

In the center of the casting is a large boss C which supports a press-fit spindle D and the turntable, which has a snug fit on a bronze bushing. The turntable holds all clamping units. Plate F has twenty accurately ground vees around its edge, one vee for each piece of work.

Each clamping unit has a hardened roller G mounted on a dowel H and is free to rotate. A roller holder J is a slide fit in a hole bored in the boring block and has a shoulder at the end for a clamping pad K which clears the face of the table. The semicircular face of the pad is accurately milled to conform to the diameter of the work while a ma-

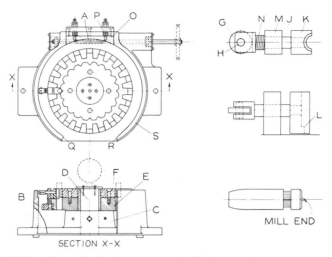

Fig. 10. Milling fixture designed to automatically clamp and release small workpieces
held by moving spring-clamp devices.

chined ledge L supports the work during the operation. Bearing block M
holds the clamping member and permits the pad to move in the direction
of the plate by the action of spring N.

On the geared end of the fixture is an adjustable pressure block O,
held in an outward position because of the pressure from springs P.
By adjusting nuts on the holding studs, the spring tension is tightened
or loosened.

In operation, the turntable rotates counter-clockwise continuously,
passing all the clamping units under a standard milling cutter. As the
roller bears on a track, machined in the bore of the casting, and the units
pass the gap in the casting between points Q and R, the spring on each
unit pushes the clamp out to its maximum extension between the pad
and the vee in the plate. The released workpiece drops clear of the
fixture into a receptacle.

When the roller on each clamping unit contacts point S, it rides along
a gradual incline on the roller track. With a clamp in this position, the
operator drops the work between the locating vee and the clamping pad,
the bottom of the work resting on the ledge. As the roller moves along
the track, the pad is pressed more firmly around the diameter of the
work. When the clamping unit reaches the pressure-block, the extra
pressure holds it tightly for deburring.

After the roller passes the pressure-block, it rides down the incline of
the track to the open gap where the spring releases the finished work.

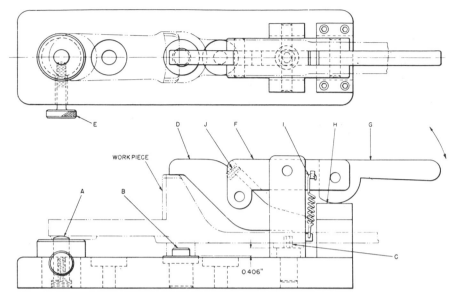

Fig. 11.   Milling fixture that employs a cam-actuated clamp to secure the workpiece in place during the final machining operation.

This operation is automatic and continuous. As the operator drops work into the clamping units, other parts are being machined.

Milling Fixture with Cam-actuated Clamp

A milling fixture designed for rapid and easy setup in machining tracer arms of engraving machines is shown in Fig. 11. The important feature of this fixture is the clamping device, which applies pressure on the work at two points by a single movement of a cam type lever. The operation performed is the finish-milling of a radial surface.

Three locating pins A, B, and C position the tracer arm in the fixture. Surrounding these pins are ground surfaces which support finish-machined bosses on the arm. A cam-actuated clamp D secures the work in place.

Pin A is a spring-loaded plunger which can be depressed and locked in place by means of a knurled-head screw E. Pin B forms part of a stepped cylindrical unit which is a press fit in a hole in the base of the fixture. A similar unit, differing only in that the pin C is diamond-shaped, is a press fit in another hole in the fixture base.

Clamp D is connected by a rocker arm F to a cam G which acts against a cam-plate H. When the lever on the end of the cam is depressed, the movement of the rocker arm causes pressure to be applied by clamp *D* to the tracer arm in two places, as seen in the illustration. For loading and unloading, the clamp is held in a withdrawn position by means of the springs *I* and *J*.

In loading, locating pin *A* is depressed and locked if not already in this position. The tracer arm is then placed in the fixture from the left and positioned on pins *B* and *C*. Pin *A* is now unlocked to permit it to rise and enter a hole in the work, which further positions it. Next, the tracer arm is clamped by moving the cam lever in a clockwise direction. The unloading is accomplished by moving the cam lever in a counterclockwise direction and depressing pin *A*. This permits the tracer arm to be lifted from the other two pins and removed from the fixture.

Quick-acting Multiple Milling Fixture

A milling fixture capable of gripping four workpieces simultaneously may be seen at X in Fig. 12. Rapid clamping, unclamping, and ejection of parts are timesaving features of this tool. The fixture is of simple construction, thereby keeping its cost at a minimum.

Four holes are drilled into a tool steel body *A* to receive workpieces *B*. Holes of smaller diameter are drilled on the same four centers to provide guides for ejector-pins *C*. Two narrow slots *D*, (view Z–Z), milled at right angles to each other, pass through the centers of these holes and are cut slightly deeper than the bottom of the larger holes. The slots impart the action of a spring collet to the upper end of the body.

External threads, which are cut around the upper part of the body, mate with the internal threads of the nut-like clamping member *E*.

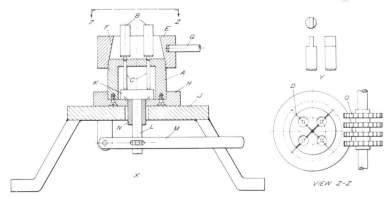

VIEW Z–Z

Fig. 12. Milling fixture that holds four workpieces simultaneously provides rapid loading, unloading, and ejection of parts.

Due to the action of the mating tapers at F, the split face of the body may be closed or opened by rotating the nut. A handle G is furnished for ease of operation.

A base H is screwed to the fixture body, and this complete unit is clamped to auxiliary table J. This table is bolted to the work-table of the machine during the milling operation. Body A is secured to its own base rather than directly to the table so that the same table may be used to hold any one of several similar fixtures.

All four ejector-pins are threaded into a single ejector-plate K, which is actuated by rod L and lever M. The rod passes through a guide bushing N that is pressed into the base. Lever L extends between two legs of the auxiliary table.

To use this fixture for machining a tongue on the cylindrical workpiece shown at Y, nut E should be backed off far enough to relieve any pressure on the tapered surface of the body. One part is then inserted in each of the four large holes. Tightening the nut will force the split gripping surface together, thus holding the parts securely.

By using a gang-milling setup O, the tongues may be finished in one pass. Part removal is very simple. Merely backing off nut E will unclamp them; then lifting the ejector lever will unseat and raise the parts for easy removal.

Shaping and Slotting Fixture for Hand Miller

Occasionally it becomes necessary to shape some pads on the inside of cast-iron box jigs. When there is no shaper available, a fixture can be

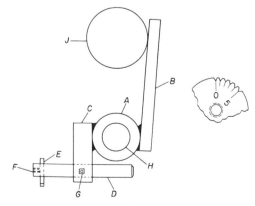

Fig. 13.   Shaper tool extension permits interior cutting and grooving of the edge of the pistol nut graduations at the right.

built up by welding, as shown at the left of Fig. 13. This consists of a piece of heavy steel tubing *A* and two pieces of steel *B* and *C*. Through bar *C* a 3/4-inch hole is reamed to accept round bar *D*. This bar is movable axially in the slot and is held by set-screw *G*. Tool *E* is held in one end of the holder by set-screw *F*.

Lately, in the manufacture of a precision sight for a pistol, some nuts were required which were graduated on the face and had corresponding grooves cut in the rim, as shown in the enlarged view (inset at the right).

The fixture has proved practical for both shaper jobs. In use, tubing *A*, which was bored to fit over 1-inch bar *H*, is clamped in place on the miller arbor. Heavy rubber bands hold bar *B* to the overarm of the miller *J*. The bands stretch, releasing tool *B* on the back strokes. Work is clamped on the miller table, and cuts are made by either moving the table or the arbor head, using their respective hand levers.

Tooling for Boring

Boring Bar for Tapered Holes

Tapered holes can be produced by the special boring bar shown in Fig. 1. The bar is equipped with a cam-guided toolslide. Designed for machining large workpieces on a horizontal boring mill, the tool is an excellent substitute for a tapered reamer, which may not be available in the exact size required for a specific job.

Principle of operation of the boring bar is similar to that of a standard engine lathe attachment in that the tool guide bar is a bar type cam. The cam is nonadjustable and revolves with the tool.

The boring bar is slotted to receive the bar cam and a cap member retains the bar cam in place. A rotating sleeve held in a steadyrest supports the outer end of the boring bar during the operation. Although the bar cam rotates, it is maintained in a fixed axial position with respect to the workpiece by means of a holding stud in an end cap attached to the rotating support sleeve.

The toolslide is keyed to the bar cam, which feeds the tool radially in accordance with the required taper angle and the axial position of the machine spindle. A different cam and toolslide combination is required for each angle of taper. The cam can be reversed to cut the same taper in the opposite direction. Matching keyways also can be generated with the same tool.

Arrangement for Taper Boring a Large Part

A large propeller, approximately 19 1/2 feet in diameter, was required to have a tapered bore. However, since a vertical lathe of sufficient size to handle the job was not available, some alternate method of machining the taper had to be developed. The solution was obtained by using the

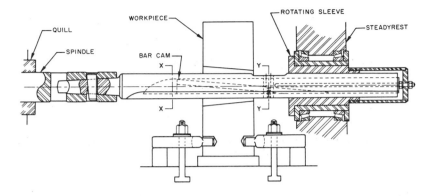

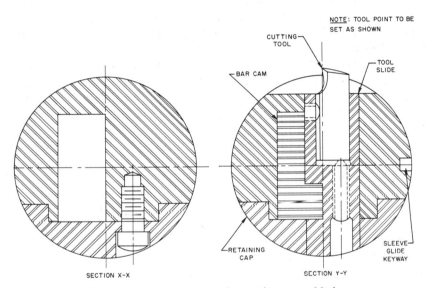

Fig. 1.   Special boring bar for cutting tapered holes.

setup shown schematically in Fig. 2 in combination with a floor type boring machine.

The device consists essentially of a hollow boring-bar *A* supported at one end in a spherical ball bearing *B* and held at the other end on a pivot shaft in a bracket *C*. The bracket is secured to the slide of the machine's facing plate *D*, and bearing *B* is held in a steadyrest. A tool-slide *E* (in the form of a sleeve) is mounted on the bar and is attached to a nut *F*, which is constructed to extend through a long slot in the boring-bar. The nut is driven by a feed-screw *G*, which is free to turn in bearings within the bar.

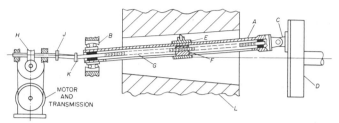

Fig. 2.   Device for taper-boring a large part. Taper angle can be adjusted by means of slide of facing plate *D*.

The feed-screw, in turn, is driven by means of a worm and gear speed-reduction unit *H* through a pair of flexible couplings *J* and *K*. The reduction unit is driven through a four-ratio transmission by a variable-speed electric motor.

Slide *E* has a groove in which the tool or a special holder is placed. The taper angle (the angle inscribed by the bar in its rotation) may be varied by adjusting the position of the facing plate slide. When this slide is adjusted, the bar is pivoted in bracket *C* on a shaft supported by two roller bearings preloaded to eliminate any backlash. The exact inclination of the bar can be determined by a precision level, a sine bar, or any other suitable gage. A clutch for disengaging the feed is built into the speed-reduction unit *H*.

In operation, the bar pivots in the fixed spherical bearing, which is centered on the axis of the boring-machine spindle. The other end of the bar is driven eccentrically by the facing plate of the machine so that the bar, in its motion, inscribes a cone having the same angle as the required taper. The tool is driven along the bar by means of the feed-screw drive system to cut the taper in the workpiece *L*. The rotation of the bar alone, however, will cause the nut to feed along the bar, the rate depending on both the lead of the screw and the speed of the boring-machine spindle. Therefore, to determine the exact feed of the tool produced by the lead-screw drive system, the feed caused by the bar rotation must be taken into account.

Adjustable Taper Boring Attachment

Figure 3 shows a simple attachment for machining large-diameter open cores in castings to tapers of various side angles. Although the device is intended primarily for horizontal boring mill applications, it is readily adaptable to a vertical boring mill or a conventional lathe having no other provision for producing internal tapers. In principle, a body fixed to and revolving with the regular boring-bar of the machine carries a slide that is adjustable to the required side angle of the taper. Feeding along the slide by means of a star-wheel is a small tool-block containing a single-point cutter.

The body A, which is a steel block trapezoidal in shape, fits closely over the boring-bar B of the machine, and receives positive endwise location by its bearing against the integral flange C of the bar. A key D and two set-screws E secure the body to the bar. To carry the slide F, a slot G is milled in the block parallel to the block axis and accommodates a rib H extending from the slide. This rib is held in the slot by a dowel-pin J, around which the slide can be swiveled to the required angle. Accurate setting is obtained by means of a fine-threaded jack-screw K engaged in a tapped hole in the tail-like end of the rib and working against the bottom of the slot. Set-screws L in the rear wall of the slot tighten against one side of the rib, locking it in position in the slot.

In a channel running the length of the slide proper is a fine-pitch lead-screw M, free to rotate but restrained from axial movement by shoulders

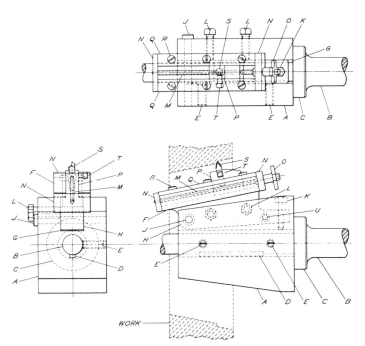

Fig. 3. This taper boring attachment revolves as a unit with the regular boring-bar of the machine, but feeds independently by means of a star-wheel.

that bear against the two end-plates *N*. Keyed to the right end of the lead-screw is the star-wheel *O*. The tool-block *P* fits the channel and has the cross-section of an inverted "T." Its base is internally threaded to form a nut for the lead-screw. Keeper plates *Q*, which are retained by screws *R*, resist any tendency of the tool-block to rise. For correct cutting action, the hole for the single-point cutter *S* is located in the tool-block so that the cutting edge is on the vertical center line of the machine spindle. A set-screw *T* holds the cutter in position.

In operation, the tool-block is retracted to the star-wheel end of the slide, and the cutter is set out radially to the diameter of the larger end of the tapered cut. A kicker bar is fastened to the frame or bed of the machine, in position where it will intercept the star-wheel, causing the lead-screw to rotate 90 degrees for each revolution of the spindle.

The length of the slide is determined by the length of the taper; it is practical to design the slide so that it is approximately 25 per cent longer, thus providing sufficient traverse for the tool-block. Where the attachment is to be used intermittently for boring duplicate parts, a register pin *U* can be inserted in a reamed hole running through the slot and rib, providing automatic adjustment to the setting. If various angle settings have to be made frequently, spacers of known thickness can be used. A bridge piece is then fastened across the top of the body, straddling the channel immediately above the tail of the rib. By inserting a spacer of predetermined thickness between the bridge piece and the tail, the slide is able to be quickly adjusted and locked at the correct setting.

Right-Angle Indexing Head for Horizontal Boring Mill

Bed plate and gib rail pads for steel press frames are face-milled on a horizontal boring, drilling, and milling machine with the attachment illustrated in Fig. 4. This device converts the drive 90 degrees, permitting

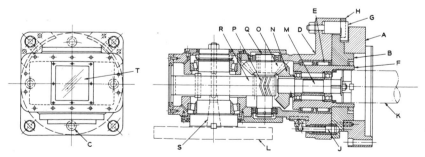

Fig. 4.   The head transmits the horizontal drive of the machine spindle *K* to the vertical plane of the face-milling cutter *L*.

the work to be clamped directly on the machine table. An integral index-ing feature enables the cutter to operate at four different points in the vertical plane.

Section and end views are shown in Fig. 4. A mounting plate *A* is doweled and screwed over the end of the machine spindle *K*. This plate is fitted with a stationary indexing ring *B* in which are four holes *C*, radially spaced 90 degrees apart. The indexing head *D* is supported on needle bearings *E* over a sleeve *F* press-fitted to, and extending from, the bore of the indexing ring.

Hook-bolt clamps *G* and retainers *H* secure the head with a shoulder on the back of the indexing ring. In the head is a cam-operated locating pin *J* which can be engaged with any of the four holes in the ring after the head has been rotated to the desired position.

From the machine spindle *K* the line of transmission to the face-milling cutter *L* is through a ball-bearing drive shaft *M*, bevel gears *N* and *O*, and pinion *P* on jack shaft *Q* to gear *R*. This gear encloses the attachment spindle *S*, which is mounted on roller bearings. Seals on the spindle and drive shaft prevent oil leakage. An inspection window *T* is provided at the end of the head.

Tool for Internal Machining of Pressure Vessels

Spherical pressure vessels up to 15 inches in diameter can be machined internally with a special bar type tool. The aluminum-alloy vessels are designed for working pressures up to 4000 psi, and are produced from flat blanks by cupping and press-swaging techniques. With this method of forming, there is a tendency for the internal surfaces in the area of the neck to deform, and in extreme cases, these surfaces may even wrinkle.

Since such deformations are undesirable, the vessels are internally machined by means of the special bar type tool illustrated in Fig. 5. The bar clears the bore of the neck and is slotted over the greatest part of its length. Housed within the slot is a swinging tool-arm which has its lower end pivoted on a pin passing through this bar. A tension spring is stretched between one side of the tool-arm and the end of the slot in the bar, so that the arm tends to take a position parallel with the longi-tudinal axis of the bar.

A pad which is free to rotate is mounted on the lower end of the bar. A member carrying a roller at its lower end is mounted so that it can slide longitudinally within the upper portion of the slot in the bar. When the bar is set up on a boring mill, it is held by an adapter sleeve in

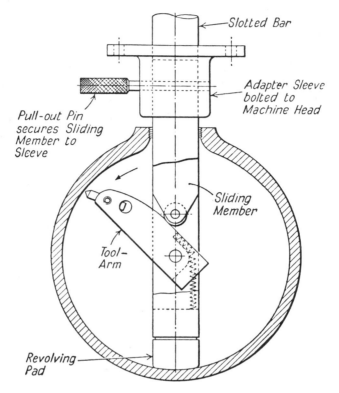

Fig. 5. A bar type tool has a tool-arm pivoting from a slotted bar, allowing a cutting
action that makes the interior surface of the sphere uniform.

which it is free to slide vertically. The inner sliding member is secured
to this sleeve by a pull-out type pin. When the bar is fed downward
through the neck opening, the pad at the lower end contacts the work
and movement is arrested. However, the sliding member carrying the
roller continues to advance. The roller then engages a cam-profile on
the inner edge of the tool-arm and swings the latter about its pivot point
so that the cutter describes an arc.

The proportions of the bar are such that the pivot point of the swinging arm coincides with the center of the spherical outer surface. A cut,
started on the thickest wall section near the neck, diminishes gradually
in depth as the tool-arm is forced downward and outward. By appropriate radial setting of the cutter, the diameter of the area so machined can
be varied as required. Also, irrespective of the area covered, the cut will
blend smoothly into the surrounding surface of the workpiece being
machined.

The work is rotated at 34 rpm and the sliding member is fed at the rate of 0.005 inch per revolution. Usually, the irregularities can be removed completely, by taking three successive cuts, each 0.03 inch deep. After each cut, the tool is reset by making a direct radial measurement from the pivotal center.

Combination Boring and Reaming Tool

Drilled or cored holes can be bored and reamed in a single operation with the combination tool shown in Fig. 6. The cutting elements are removable so that the tool can be adapted for operating in a wide size range.

Body *A* is joined to tapered shank *B* by ring-nut *C*. Tang *D* on the front of the shank transmits a positive drive through its fit in a slot in

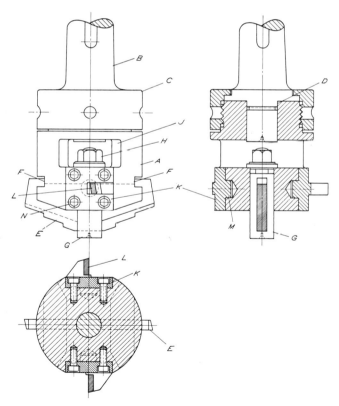

Fig. 6.   Spade type stepped cutter *E* and single-point cutters *L* bore and ream at the same time.

the body. For boring, spade type stepped cutter *E* is retained in a diagonal slot in the end of the body. Shoulders *F* centralize the cutter with the body, and threaded yoke *G* encloses the end of the cutter and secures it in position through engagement with nut *H*. Milled area *J* is made large enough to accommodate a wrench for the nut.

One or two holders *K*, having carbide cutter tips *L*, can be set in the body for reaming. The holders are aligned in the body by pin *M* and are secured by set-screws *N*.

Swiveling Tool-Holder for Cutting Oil-Grooves in Bushings

Tools held in a boring-bar in the conventional manner and employed for cutting oil-grooves in medium-size bushings were found to have a very short life. The operation was performed on a horizontal boring machine equipped with a grooving attachment. Figure 7 shows a development of a double oil-groove cut in a bushing 7 5/8 inches in diameter, which is typical of the work handled. When one of the endless oil-grooves had been cut, the bushing was indexed through an angle of 180 degrees, and the second groove shown was generated.

Since the tool, which was rigidly held in the boring-bar perpendicular to the bore of the bushing, had to be fed forward and then backward an equal amount during each revolution of the boring-bar, it had to be ground so that it could cut in both directions. Also, it was necessary to provide excessive side clearances on the tool to prevent interference because of the steep helix angles of the groove. The tool was therefore weakened and rapidly failed. Furthermore, the grooves produced in this way varied in width.

To overcome these faults, the tool-holder shown in Fig. 8 was designed. This holder is free to swivel in a vertical plane, so that the top cutting face of the tool bit will tend to assume a position perpendicular

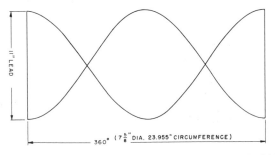

Fig. 7.   Development of a double endless oil-groove having a lead of 11 inches, which is cut in the bore of a bushing 7 5/8 inches in diameter.

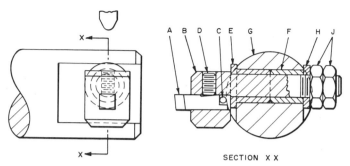

Fig. 8.　Details of a swiveling tool-holder which permits the top cutting face of tool bit *A* to remain perpendicular to its path of travel.

to its helical path of travel. In this way, grooves of uniform width are generated and a much longer tool life is obtained.

Tool bit *A* is located in holder *B* against stop-pin *C*, and is secured by a hexagon-socket set-screw *D*. The shank of the holder is a snug sliding fit in slip bushing *E* and liner bushing *F*, which are pressed into a cross-hole in boring-bar *G*. While the holder is free to swivel, it is prevented from moving axially by thrust washer *H* and lock-nuts *J*, which are screwed on the threaded end of the holder.

It is necessary that the tool bit be ground with minimum front and side clearance angles and no top side rake. The top cutting face of the tool is aligned with the center of the boring-bar, while the axis of the tool-holder shank is offset. The holder is, therefore, forced to swivel and keep the tool aligned with its path of travel.

General Operation Tooling for Lathes and Turret Lathes

Pneumatic Lathe Mandrel Compensates for Component Variations

A cylindrical sand casting, such as the one shown at X in Fig. 1, required machining to bring the periphery concentric with the cored cavity. The cavity is circular in shape at the open end, changing to a smaller circular shape having two opposing flats at the closed end. Due

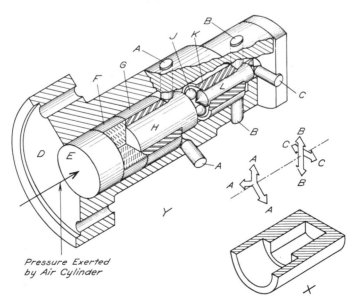

Fig. 1. Equalizing plungers in air-actuated mandrel will grip within rough castings regardless of variations from one cored hole to the next.

to variations in the dimensions of the rough-cored hole, concentricity could not be maintained by locating from fixed mandrel points.

Consequently, the pneumatically operated mandrel at Y was designed to provide a three-point, self-centering arrangement A for the open end of the cavity, and two sets of equalizing plungers B and C for the closed end. Compensation has been provided to permit plunger conformity to shape variations of the cored holes.

Mandrel body D is bolted to the spindle nose of the lathe. The interior of the mandrel body is bored out to receive the clamping mechanism which is operated by an air-cylinder push-rod exerting pressure on plug E. A close-fitting rubber plug F serves as a hydraulic medium against the ends of sleeve G and plunger H. The diameters of these two parts are proportioned so that their end areas are approximately equal, allowing even distribution of pressure from plug F.

On its forward end, sleeve G has a chamfer that coincides with the pointed end of clamping plungers A. When the sleeve is advanced, all three clamping plungers are forced out radially an equal amount.

The forward flat end of plunger H presses against three steel balls J. These balls lie in an annular groove formed by an internal chamfer on sleeve K and the correspondingly chamfered end of plunger L. Relative longitudinal displacement of up to 1.4 times the ball diameter can be had by using a 90-degree included angle in the annular groove. The forward end of sleeve K has an external chamfer that operates clamping plungers B. Clamping plungers C are controlled by the cone on the forward end of plunger L.

In practice, a suitable means should be provided for retaining plungers A, B, and C within the mandrel body to prevent their flying out if the lathe is started with no workpiece in place. This may be a spring or a nylon plug extending from the side of the plunger or a limiting pin engaging a slot in the plunger.

Novel Work-Holder for Lathe Operations

Frequently, a flat workpiece must be turned on one end and faced to obtain an accurate dimension. A good device for holding such parts in a collet or chuck for these operations is illustrated in Fig. 2.

The holder consists of two clamping members A and B which are hinged on dowel-pin C. The part D is set in a nest in member B, as shown in section Y-Y and is positioned by diamond locator pin E. In this case, locating the workpiece by the 5/16-inch-diameter hole permits the 1.3125-inch dimension to be accurately held. Member A serves only as a clamp. Clamps A and B are assembled with the hinge pin and the

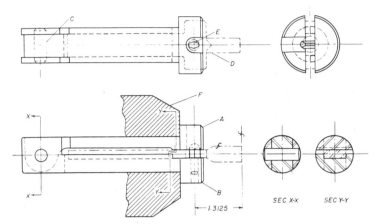

Fig. 2. Hinged work-holding fixture that fits in chuck or collet to facilitate performance of lathe operations on flat parts. Component is positioned by a diamond locator pin.

assembly is ground on the periphery. This makes it possible to hold the assembly in a collet *F* (or chuck) for a turning or facing operation.

Quick-releasing Chuck for Threaded Parts

Owing to a change in the design of a certain recording instrument, a large number of studs of the type illustrated at X (Fig. 3), had to be modified. The stud, made from medium-hard brass rod, consists of two threaded shanks of slightly different diameter, pitch, and length. Both shanks are threaded practically full length, almost up to the end faces of a short conical shoulder, as shown. All portions are concentric, and the surface of the conical shoulder is ground to a smooth and accurate finish.

The changes required remachining of the extreme end of the smaller diameter shank to remove the threads and produce a short plain cy-

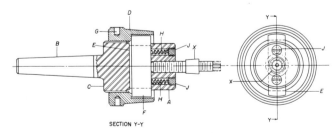

Fig. 3. This special chuck for holding a threaded stud *X* is designed for quick release of the workpiece.

lindrical portion less than half the diameter of the original shank. This portion had to be held to close tolerances on diameter, parallelism, concentricity with the remainder of the shank, and axial alignment with the entire stud. The stock to be removed is indicated on the drawing by broken-line cross hatching.

For the turning operation it was found impracticable to grip the studs in a three-jaw self-centering chuck or a collet, owing to the risk of damaging the precision threads and producing eccentricity in the finished work. Gripping the studs in a draw-in collet could cause an additional difficulty. Variations in the amount of draw-in movement when closing the collet sleeve could result in a difference in the machined lengths of the parts. To overcome these objections an auxiliary chuck of unusual yet simple design was successfully used to hold the studs by the threads on the large-diameter end. The chuck was designed to insure rapid release of the finished stud and prevent damage to the threads.

As seen in Fig. 3, the chuck consists of a cylindrical steel body A machined at one end for an integral Morse taper shank B which provides a means of accurately mounting the chuck into the nose of the live spindle of the lathe to be used. The large-diameter portion of the body, seen at C, is machined with a fine-pitch thread to receive the hardened steel adjusting nut D. This member should be fitted to screw easily (by finger pressure) over the body. A threaded hole is provided in the right-hand end of the body to receive the threaded shank of the large-diameter end of stud X. The threads are machined concentric in the body to insure that the fitted stud will run true.

A rectangular slot E is machined diametrically through the unthreaded forward part of the body, and hardened and ground steel slide F is then fitted in the slot. This member should slide smoothly, its thickness being about twice the outside diameter of the end of the stud which screws into the body. The width of the slide is about 0.040 inch less than that of the slot, thus allowing the slide that amount of free movement. The threaded hole for stud X enters slot E, its length being about 1/4 inch shorter than that of the threaded portion of the stud. The inner end of the mounted stud will thus bear positively against the right-hand side of slide F, preventing jamming between the end of the stud threads and the hole in the body.

Slide F is about 5/8 inch longer than the outside diameter of the body at slot E. Thus, each end of the slide will project about 5/16 inch beyond the periphery of the body. A deep concentric recess is machined in nut D to receive the projecting ends of the slide. For clearance purposes, the diameter of the recess is made approximately 0.025 inch larger than the length of the slide. The end faces of the slide are also curved to insure that amount of clearance. The internal shoulder at the base of the recess bears against the left-hand side of the projecting portions of the slide. Therefore, when the nut is tightened, the slide will be moved across slot E until it bears against the opposite side of the slot. The nut can be locked tightly with a spanner wrench by means of holes G.

When nut D is released, the slide is instantly retracted by means of the two light compression springs H placed in holes drilled through the right-hand end of the body into slot E. These springs are retained by fixed plugs J screwed into the outer tapped ends of the holes.

In use, the chuck is first mounted into the live spindle of the lathe and checked for trueness. Nut D is then turned the maximum amount toward the right in order to push slide F against the right-hand side of slot E, and is tightened in this setting by a spanner wrench. A stud X is next threaded into the body until its end face bears against the hardened slide. The required cylindrical portion is then turned on the projecting shank of the stud, the correct length being determined by a fixed stop on the lathe slide. As each part is located positively in the same longitudinal position within the chuck, the length of turned portion is precisely the same on all studs, and the lathe slide stop does not require readjustment.

After the stud has been machined, it is instantly released by simply unscrewing nut D toward the left. This allows the internal slide to move in the same direction under the action of springs H. The finished stud can then be quickly unscrewed by finger pressure. With a chuck of this design, brass studs were economically modified to the required close tolerance, concentricity, and axial alignment without inflicting damage to threads or the finish ground shoulder.

Mandrel Expanded from Revolving Center

An expanding stub mandrel, controlled from a revolving center, cuts loading and unloading time for secondary lathe operations. It has wide application; the principal requirement is that the work have a finished bore and one finished face.

The device is shown in Fig. 4. At its left end, the mandrel corresponds in design to the lathe spindle. In this instance, the mandrel is threaded externally and tapered internally for an American Standard spindle. The right end of the mandrel is reduced in diameter to fit the work bore. This surface is cylindrical, except for its very end, which is beveled slightly to aid in loading the work.

There are three slits through the body of the mandrel, extending from the right end to the center. These are spaced radially at 120 degrees and

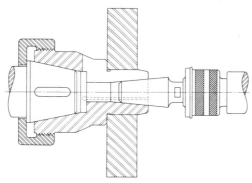

Fig. 4. The work is secured to the mandrel by the thrust of the expander plug carried by the revolving center.

provide a necessary amount of flexure. Internally, the right end is bored to a slow taper to receive an expander plug carried by a revolving center held in the tailstock spindle.

Several manufacturers of ball-bearing revolving centers design them with a removable center point, usually threaded to the rotating part of the tool. Wrenching flats on the point provide easy removal. The expander plug has a threaded shank, corresponding to the thread in the rotating part, so that it is readily interchangeable with the regular center point. All bearing surfaces of the stub mandrel and expander plug are hardened and ground.

To grip the work, the tailstock spindle is advanced and locked, with the expander tightly in the end of the mandrel. For loading and unloading, the tailstock spindle is unlocked and retracted.

Quick-acting Chuck for Tapered Parts

The quick-acting chuck illustrated in Fig. 5 was designed for holding tapered parts. The workpiece shown required a secondary machining operation to produce the step in the small-diameter end. This step was omitted through oversight in the original machining operation. The over-all length of the part had to be maintained within close tolerances, and the end surfaces were required to be as flat as possible and square with the longitudinal axis of the part. Because of the application of the part, the end surfaces could not be center-drilled for mounting between centers in the usual way.

The body of the chuck is machined at one end to form a parallel or tapered shank, which allows the chuck to be gripped in the collet of the lathe. A centering plug is screwed into the shank end of the body. This

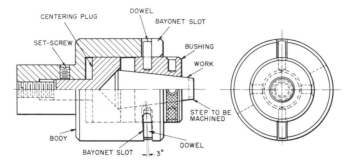

Fig. 5. Quick-acting chuck that exerts a powerful and positive locking action on tapered parts with a minimum of effort. Precision tolerances are maintained, due to the fact that consistent end-wise location is insured.

plug has a conical recess machined in the face on the flanged end to suit the chamfer previously produced on the work. The conical recessed surface of the plug should be hardened and ground to minimize wear and insure accuracy in centering the tapered part.

A small screwdriver slot is machined across the left-hand end of the centering plug to facilitate adjusting the plug within the body. The plug can be locked in any desired position by means of a brass set-screw that bears on the unthreaded portion of the plug shank.

The hardened brass locating bushing is a snug sliding fit in the large bore at the front of the chuck body, and is taper-bored to fit the work-piece. Two hardened steel dowel-pins, driven into diametrically opposite holes in the bushing, project radially outward through a pair of bayonet slots machined in opposite sides of the body. The bayonet slots are made about 1/64 inch wider than the diameter of the dowel-pins, and are inclined around the periphery of the body at an angle of 3 degrees.

Entrances to each slot extend out through the end of the body, and are chamfered to allow the dowel-pins to enter readily. The portion of the bushing that normally projects from the end of the body is knurled to facilitate gripping by hand, and is also provided with three or four radial holes to permit the use of a spanner wrench or drill rod when loading or removing the bushing.

In operation, the tapered workpiece is inserted into the bore of the bushing, and the loaded bushing is slid into the body of the chuck. When the dowel-pins engage the bayonet slot, the bushing is rotated by hand. The inclination of the slots will cause the bushing to be drawn gradually into the body.

When the chamfer on the work contacts the conical recess in the plug, the workpiece will automatically and accurately become centralized with the chuck body. By then applying a spanner wrench to the bushing, sufficient pressure can be exerted to hold the work securely during machining. Overhang of the part beyond the end of the chuck can be varied by loosening the set-screw and adjusting the centering plug with a screwdriver.

With this simple, low-cost chuck, a powerful and positive locking action is obtained quickly by a minimum of effort. Precision tolerances can be maintained, and the finish previously applied to the tapered surfaces of the parts is not damaged. Each part inserted into the chuck will occupy the same endwise position, regardless of variations in the over-all length of the parts. As a result, once the cutting tool has been set up, each part may be faced to exactly the same length.

By providing a number of locating bushings, each having a different size tapered bore, one chuck can be used for many different parts. The shape of the bore can be varied to handle parts other than conical. Also, the inclination angle of the bayonet slots can be changed to alter the amount of locking movement and the pressure applied on the work.

Lathe Chuck Useful in Holding Non-cylindrical Round Work

An effective type of universal lathe chuck for holding short tapered parts — for example, tapered dowel pins or short non-cylindrical (but symmetrical) round parts such as machine handles — is illustrated in Figs. 6 and 7. Operations like end facing and end drilling can be readily performed, with the advantage that the chuck is immediately and automatically adjustable to any degree of taper or variety of contour of any object within its capacity. With work of this nature held in a conventional chuck, it is necessary to use a set of soft jaws or an intermediate split bushing, either of which must first be bored to correspond to the contour of the work. What is more, the soft jaws or split bushing must be rebored for each application.

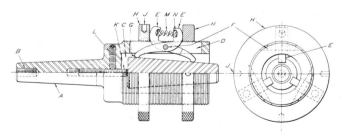

Fig. 6.   The tapered object in the chuck is automatically centralized and gripped when the three pairs of arms *E* move radially in unison.

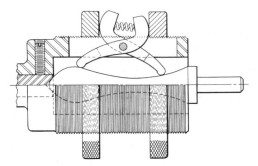

Fig. 7.   The arms adjust themselves with equal facility to a non-cylindrical, but symmetrical, round object such as this machine handle.

In the half-sectional view of the chuck, Fig. 6, are shown construction
details.   One end of its steel body is reduced to form the shank *A* that
conforms to the taper of the spindle hole of the lathe.   A tapped hole *B*
receives a draw-in bar that maintains the chuck in position.   The body
of the chuck is bored out at *C* to a diameter that is great enough to receive the size of objects for which the device is intended.

Three slots *D* are milled radially into the body, spaced 120 degrees
apart.   Within each slot is a pair of hardened steel arms *E*.   These are
pivoted on a fulcrum pin *F* pressed in a hole cross-drilled through the
sides of the slot.   The arms in each pair are stepped down to half thickness in the area of the pin so that they are in line.   The longer part of
each arm is within the bore *C*, with its extremity rounded to form a
gripping point *G*.   The shorter and upper part of the two arms *E* bears
against the clamping nuts *H* engaging the exterior of the body, which is
threaded.   It will be noted that the upper edges of the arms are cambered
for smooth action against the faces of the nuts.

For ease in manually adjusting the nuts, their periphery is knurled.
Each nut also has several small holes *J* drilled radially to receive a spanner wrench used in applying locking pressure when a workpiece is
being gripped.   All three pairs of arms must be made exactly the same
shape and size.   A good way to attain the necessary precision is to machine a bar (of sufficient length to make all six arms) to the required
outline shape, and to drill the bar for the fulcrum pin-hole.   The bar can
then be sawed to produce six arms of the proper thickness.   These require
stepping to half thickness on one side in the area around the fulcrum
pin hole.

In many applications, the over-all length of the object is critical.
For this purpose, a threaded stop-rod *K* can be extended a suitable
distance into the bore *C*.   A set-screw *L* maintains the setting of the

stop-rod. When positioning an object, its rear face is held firmly against the stop-rod before any locking pressure is applied.

Between the upper parts of each pair of arms is a light compression spring *M*, both ends of which are contained in shallow blind holes *N*. The purpose of this spring is to spread the arms apart when the nuts *H* are released, thereby facilitating unloading and reloading.

After an object has been positioned against the stop-rod, both nuts are screwed toward each other until they are hand-tight, then each is locked with the spanner wrench. In tightening, the lower parts of the arms converge toward the center line of the chuck, and since the three pairs of arms move in unison, the object is automatically centralized.

The larger diameter of the object which the chuck holds in Fig. 6 is toward the back of the body. However, the object can be reversed in the chuck, and the arms will automatically readjust themselves. Figure 7 shows how conveniently a machine handle was held in the chuck when it was necessary to recut the cylindrical shank at one end. It will be noted that the stop-rod was given a conical seat to provide a suitable endwise location for the handle.

Expanding Arbors Utilize Rubber Elements

The fact that pure gum rubber is sufficiently fluid to flow, and at the same time not compressible, makes it an ideal material for use in expanding arbors. Such arbors are easy to make, run true, and grip effectively. One style, shown in Fig. 8, consists of a series of rubber bushings *A* which are located between steel spacers *B*. A split sleeve *C* encloses the bushings and spacers.

The work *D* has a slip fit over the sleeve, and is located endwise against an integral shoulder *E* of the arbor. When the nut *F* is tightened, pressure is exerted on the faces of the rubber bushings and transmitted to the inside diameter of the sleeve. In constructing this arbor, the bushings and spacers are ground to size in position, with the nut tightened just enough to cause the bushings to bulge slightly, as in Fig. 9.

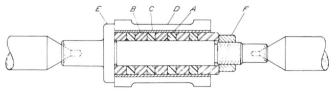

Fig. 8.   Tightening the arbor nut *F* causes the rubber bushings *A* to expand against the inside diameter of the split sleeve *C*.

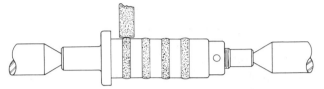

Fig. 9.   In grinding the bushings and spacers to size, the arbor net is tightened sufficiently to produce a slight bulge of the rubber bushings.

For holding small work, the arbor can be designed without the split
sleeve, in which case the bushings expand directly against the inside
diameter of the work. The rubber used should have a Durometer value
of 60 to 75. It is advisable to take light grinding cuts to avoid chatter
marks in the bushings, and also to regrind the bushings periodically

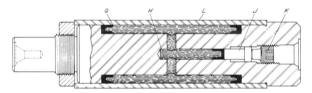

Fig. 10.   The set-screw *K* has to be turned only slightly for the rubber to press against
the split sleeve *L*.

because of the frictional wear developed in loading and unloading
the work.

An exceedingly accurate arbor uses a somewhat more plastic form of
rubber in a simple hydraulic system. In this design, seen in Fig. 10,
the rubber occupies an annular area *G* around the body of the arbor, and
also a feeder area *H* which extends to a plunger *J*. By tightening set-

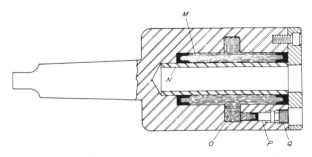

Fig. 11.   Split sleeves of different internal diameter can be readily interchanged to
increase the holding range of the chuck.

screw *K*, the plunger is advanced, forcing the rubber to expand around the inside diameter of the split sleeve *L*. This arbor produces an effective grip where the tolerance of the hole in the work is held within 0.003 inch. Very little runout is experienced.

A similar principle is embodied in the design of the chuck, Fig. 11, for holding work externally. Here, the rubber occupies an annular area *M* around a split sleeve *N* that lines the bore of the chuck. Rubber in the feeder area *O* extends to a plunger *P* which can be operated by a set-screw *Q* that is accessible from the face of the chuck.

Hydraulically Operated Finger Chuck

A finger chuck, designed for use in boring and grinding operations where close tolerances must be maintained, is shown in Fig. 12. A unique feature of this chuck is its method of clamping; it has a self-contained hydraulic system.

Chuck body *A* is drilled and counterbored in two places for mounting on a conventional lathe faceplate. The front face of the chuck is bored and counterbored to receive an adapter *B* which is held in place by two screws. This adapter can be replaced with other adapters so that different work shapes can be accommodated.

The workpiece is locked in the adapter by two clamping fingers *C* that are mounted in opposing slots in the periphery of the chuck body. These clamping fingers can be interchanged with others to hold different

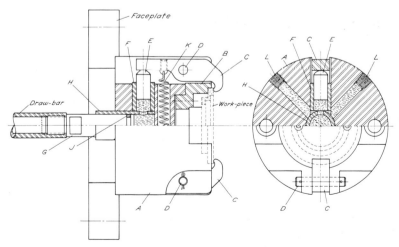

Fig. 12. Finger chuck, having a self-contained hydraulic system, is actuated by lathe draw-bar.

work diameters. Pin-shafts D, held in place by cotter-pins at both ends, support the clamping fingers so that they may be pivoted.

Locking motion is imparted to each finger by individual pistons E riding in liners F (one one of each shown). Heavy-weight hydraulic fluid is contained in passages within the chuck body itself. Pressure is applied to the fluid by the advancement of main piston G fitting within cylinder H. An "O" ring packing J prevents the bypassing of fluid to the rear of the chuck.

In use, the workpiece is first located in the adapter. Following this, the lathe draw-bar is moved toward the headstock. Main piston G, which is threaded into the end of the draw-bar, is consequently forced into cylinder H, tending to compress the hydraulic fluid. The increased pressure within the chuck forces both pistons E outward into contact with the under side of the clamping fingers. These fingers, which are pivotally mounted, grip the workpiece tightly.

To release the clamping fingers, the draw-bar is moved away from the lathe headstock, taking the main piston with it. Pressure behind the two smaller pistons is decreased, permitting tension springs K to return the fingers to their original position.

Filling the chuck with hydraulic fluid is accomplished by first withdrawing piston G to its rearward limit of travel. With both plugs L removed, fluid may be funneled into one opening while the other opening serves as an air bleed. After the plugs have been replaced, the unit is ready for use.

Centering and Side-clamping Collet

The collet shown in Fig. 13 was designed to locate the workpiece E, first against a shoulder on sleeve D and then by the slotted expanding collet sleeve G in order to clamp it in a central position. This collet can be used when a hole or a number of holes have to be drilled accurately at a distance a, for example, from one side of a ring, and also when they must be accurately positioned on a radial line or at a specified off-center distance.

Positioning of a groove at a precise distance b from the base or face of workpiece E and turning it concentric with the inside diameter can also be easily accomplished when using this collet for turning operations.

The collet provides two clamping actions through one movement of the draw-rod A. A number of fingers N press the workpiece E down, and the collet sleeve G grips it centrally. To enable quick unloading of the finished parts, the fingers N must retract into cap K.

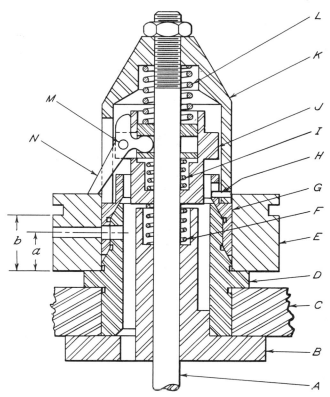

Fig. 13.   Collet designed to permit work *E* to be clamped in a centralized position by means of draw-rod *A*. This view shows collet in fully closed position.

Half-section views A, B, and C, Fig. 14, show the collet in the open, partly closed, and nearly closed positions, respectively, the completely closed position being shown in Fig. 13. In the open position, view A, Fig. 14, the upper spring *L*, Fig. 13, and the lower spring *F* are relieved, this allows the center spring *I* to expand and move the fingers *N* to the retracted position shown. In this position, the workpiece *E* can be slid off or on over the cap *K*. During the opening or upward movement of rod *A*, pins *H* pull the collet sleeve *G* upward and thus relieve the centralizing clamping pressure exerted on the work.

In the partly closed position shown at B, Fig. 14, the cap *K* moves down distance *c*, which causes pressure on all three springs, *F*, *I*, and *L*. The deflection of the springs *F* and *L* is very small, while spring *I* is compressed sufficiently to cause the fingers *N* to swing outward.

Further closing of the cap *K* to a distance *d*, indicated in view C, Fig. 14, causes the fingers *N* to clamp the workpiece *E* against the

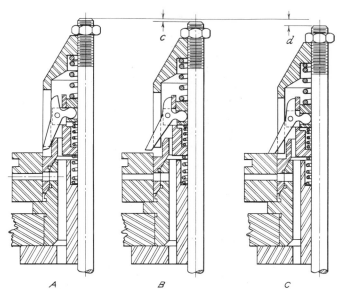

Fig. 14. (A) Half-section view of collet in open position. (B) Collet in partly closed position with hold-down fingers expanded. (C) Collet in the nearly closed position with clamping fingers *N*, Fig. 13, in contact with work *E*.

shoulder on collar *D*. The fingers *N* stay in the same position as that shown in view B because they rest against the upper part of sleeve *J*. This also causes the sleeve *J* and the fingers *N* to move as one unit.

When the cap is further closed, the spring *L* is compressed and more pressure is applied through the fingers *N* to the workpiece *E*, until the bottom of the cap *K* presses on the shoulder of the collet sleeve *G*, as shown in Fig. 13.

This collet can be mounted in any desired position through plate *C*. Ample space is provided for chips between members *B* and *D*. When the collet is used in a vertical position and a side drilling operation is performed, the chips will fall down through openings in the collar of member *B*. The collet members must be so assembled that there will be no rotary movement between members *G* and *D*, in order to keep the clearance hole on the collet sleeve *G* in the correct position for drilling.

Cam-operated Chuck Featuring Self-adjustment

A three-jaw, cam-operated chuck with self-adjusting features, made to accommodate workpieces having variations in their outside diameters, is shown in Fig. 15. This chuck is being used on a special-purpose automatic lathe equipped with a magazine feed. The workpiece is a small iron casting which has to be drilled, bored, and reamed, as well as rough- and finish-faced at both the front and the rear. The rear facing operation made it impossible to use a conventional type of air or hydraulically operated chuck having a center draw-bar. The castings are loaded into the chuck by a pusher carried in the machine turret. Ejection upon completion of the operations is effected by a forward movement of the tool-spindle used in rear facing.

There are three jaws *A* which slide radially in guide ways in flange *B* having tapered sides. The flange, which is an extension of spindle *H*, is held by screws in body *C*. The jaws are moved outward by a single-coil spring *D*, which is anchored in one of the jaws, as shown. Each jaw is provided, at its outer end, with a tapered face which is engaged by a wedge, as at *E*. These wedges slide in an axial direction along studs *F*, which are carried by the chuck-operating sleeve *G*. This sleeve is keyed so that it can slide axially along the machine spindle *H*.

Above each wedge unit is a pressure-plate *J*, which is contacted by three toggle-levers *K*. These levers are pivoted in the chuck body *C*, and their outer ends bear on the cam face of sleeve *G*, as shown. The spindle flange is cut away at three positions to provide clearance for the wedges *E*.

Opening and closing of the chuck are controlled by a camshaft at the front of the machine. Movement of the sleeve *G* to the right, to close the chuck, causes the wedges to move forward under the action of the springs *L*, until each jaw makes contact with the workpiece. Continued movement of the sleeve then results in the toggle-levers being actuated to lock the chuck, the heads of the studs *F* moving clear of the wedges, as seen in the view on the right.

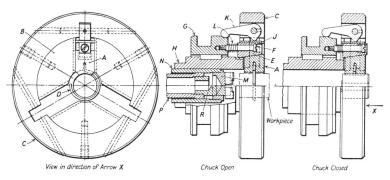

View in direction of Arrow *X*   Chuck Open   Chuck Closed

Fig. 15.   Self-adjusting, cam-operated chuck, designed to hold and eject workpieces of different diameters.

The tool for rough-facing the back surface of the casting is carried in a holder *M*, screwed into a cam-operated sleeve *N*. Within this sleeve is another cam-operated sleeve *P* which carries the finish-facing tool-holder *R*. As mentioned, the finished piece is ejected by an additional forward movement of the rear facing tool.

Chucks for Holding Workpieces by Threaded Portions

It is sometimes desirable to chuck a workpiece on a threaded portion of the part. If such a workpiece has a hexagonal or some other flat-sided surface, it can easily be unscrewed with a wrench at the conclusion of the operation. A wrench is generally necessary to break the seal formed between the work and the chuck face by the pressure of the cutting tools. If the part to be machined is round, with no flat surfaces, or if the surfaces of the workpiece should not be marked or damaged, the chuck must be designed so that this seal can be broken without using a wrench.

The chuck illustrated in Fig. 16 is used for holding a symmetrical-shaped piston by means of an internal thread while machining the opposite end of the part. The chuck seen in Fig. 17 is employed for holding a bushing by means of an external thread while finish-boring the work. An entirely different approach is necessary in designing chucks to hold workpieces by external threads than that required for chucking on internal threads.

The male chuck, Fig. 16, consists of a body which is internally threaded to fit the spindle nose of a screw machine. A stud is threaded

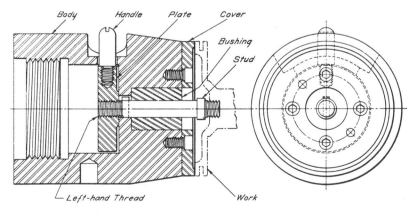

Fig. 16.   Seal formed between the work and cover by the pressure of the cutting tools is broken by moving the handle on this special chuck.

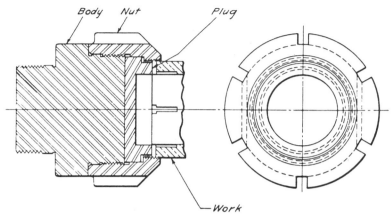

Fig. 17.   When the workpiece has been bored, the net is rotated slightly to break the seal between the work and chuck face produced by the cutting pressure.

on one end to fit the workpiece, and has a left-hand thread on the opposite end to fit a plate. The stud is a slip fit in a hardened bushing, which is pressed into the chuck body, and is also keyed to the body by means of a pin which slides in a keyway in the body.

For additional accuracy in locating the work, a hardened cover, which is screwed and doweled to the chuck body, is machined to fit the bead on the face of the piston. A handle, screwed into the plate, can be moved through an angle of 60 degrees in a clearance slot provided for it in the chuck body. The handle is pushed against one side of the clearance slot before a piece is screwed on the chuck. After the machining operation is finished, the handle is moved to the opposite side of the slot. This rotates the plate on the stud, thereby breaking the seal between the piston and the cover. A hole drilled radially in the periphery of the chuck body provides a means for mounting the chuck and removing it from the spindle nose of the machine with a spanner wrench.

The female chuck, Fig. 17, consists of a body which is threaded to fit a master chuck adapter. A nut, screwed on the chuck body, is internally threaded to fit the work. The outside of the nut is slotted to permit using a spanner wrench. The loose plug is made a slip fit in the nut, and rests against the face of the body. A recess is provided in the center of the plug to clear the boring tool.

In chucking, the work is screwed against this plug. After the workpiece has been machined, a slight turn of the nut will break the seal formed between the work and the plug. Flats are provided on the chuck body for mounting and removing the chuck from the adapter.

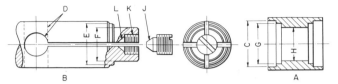

Fig. 18.   This simple expansion arbor assures the concentricity of the outside diameter
with the bore of the bushing.

As can be readily seen, these chucks can be designed to fit a large
variety of workpieces and can be used on many different types of
machines. They are economical to make, and yet provide a quick and
accurate way to hold work during machining.

Expansion Arbor for Secondary Lathe Operations

A simple expansion arbor is a popular means of supporting a bushing
when it is necessary to finish-turn an outside diameter to a high degree
of concentricity with a previously completed bore. Figure 18 shows a
typical bushing A and the arbor B which was used for the secondary
operation. A short length of steel bar having a diameter somewhat less
than that of the counterbore C is used for the arbor.

After two holes D are cross-drilled through the bar at right angles to
each other, the two diameters E and F are turned. Diameter E is made
somewhat less than counterbore G, and diameter F is roughed out to
about 1/64 inch more than the diameter of the bore H.

A standard set-screw having a dog point is beveled as at J, and can be
received in the tapped hole K in the end of the arbor. (In preparing the
hole for the tap, the included angle of the tap drill should correspond
to that of the bevel on the dog point.) As a next step, the body of the
arbor is slotted lengthwise into four sectors, from the end to the two
holes D. The slots are cut with a hacksaw at points on the diameter
where they will intersect the holes.

It is now necessary to grip the arbor firmly in a collet chuck on the
lathe in which the secondary operation is to be performed. The arbor
should be extended so that the holes D are not within the collet; otherwise, the purpose in slotting the arbor to obtain a spring action will be
defeated. With the set-screw hand-tight in the end of the arbor, diameter
F is turned to a sliding fit with bore H. As a final step, the diameters of
the arbor should be chamfered.

In operation, the bushings are slid on the arbor, after which the
set-screw is tightened with a screwdriver or Allen wrench, as the case

may be. The thrust of the bevel of the dog point against the conical surface *L* left by the top drill expands the diameter *F* sufficiently to hold the bushing securely. After the outside diameter of the bushing has been turned, the set-screw is loosened. The arbor springs back to its normal diameter, and the bushing can then be removed.

Eccentric Adapter for Lathe Chuck

An adjustable adapter designed for use with a chuck permits either concentric or eccentric turning. The adapter, which is shown in Fig. 19, is a movable member that is sandwiched between a standard lathe chuck and faceplate.

Ring *A* is doweled to a rectangular movable plate *B*. Four screws *C* pass through both the ring and the plate from the rear and engage the existing threaded holes in the back face of the chuck (shown in broken lines). The outside diameter of the ring is calculated to provide a good fit in the back recess of a chuck. A clearance hole is bored through the center of the plate to facilitate the gripping of long workpieces.

Two bolts *D* pass through elongated holes *E* in plate *B* and are threaded into faceplate *F*. The shoulder portions of the bolts that contact the elongated holes are ground and serve as guide pins. A third elongated hole *G*, lying at a right angle to the first two, receives eccentric stud *H*. Movement of the stud will force plate *B* either to the left or to the right

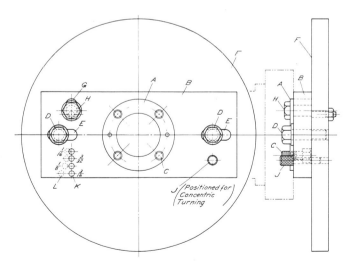

Fig. 19. Adjustable adapter fits between standard lathe face-plate and chuck to permit eccentric turning of workpieces.

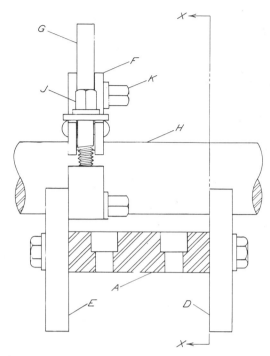

Fig. 20.   Clamping arrangement that holds bar stock to a lathe compound for machining purposes.

to effect the desired adjustment. A standard hexagonal head is provided on the stud to simplify adjustment.

When movable plate *B* is positioned so that ring *A*, and therefore the chuck, is aligned with the lathe spindle, locating pin *J* can be inserted into a hole in the plate and will pass into a corresponding hole drilled into the faceplate. After tightening bolts *D*, the device can be used for conventional turning. Adjustment of the movable plate for eccentric turning when any of four commonly used offsets — 1/16, 3/32, 1/8, or 3/16 inch — are required can be accomplished by means of locating holes *K*.

Drilled into the faceplate, and on the same horizontal center line as the locating holes, are four corresponding holes *L*. Each of the holes *L* is spaced so that, when aligned with its respective hole in plate *B*, the chuck will be located off-center by the distance stamped on the plate. These distances can, of course, be chosen to suit the particular needs of the individual shop. To insure proper alignment of the holes, locating pin *J* is inserted. Bolts *D* can then be tightened to secure the position.

Offset adjustments are not limited to the fixed spacing of holes *L*. Plate *B* can be set at any position within the limits of slots *E*. Counter-weights can be attached to the faceplate for high-speed operation.

Bar-Stock Clamp for Lathe Compound

A device for clamping workpieces made of bar stock to the compound of a lathe for drilling, milling, or facing operations is illustrated in Fig. 20. Easily adjusted to hold round, square, hexagonal-, or octagonal-shaped bars in a wide range of sizes, this versatile arrangement facilitates the machining of both production and short-run work. Clamping is accomplished by the tightening of a single bolt.

The base *A* of the attachment (Figs. 20 and 21) is mounted on the lathe compound *B* and held in place by two cap-screws. These screws are tightened into threaded holes in member *C*, which has been machined to fit the T-slot normally used for the toolpost. Work-holders *D* and *E* are fastened to the base with cap-screws that go through slots in these members. Both work-holders have a V-groove machined at both top

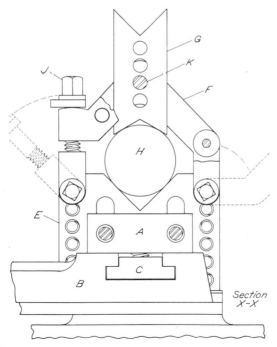

Fig. 21.   Screw *J* and clamp *F* both swing clear to allow easy loading and unloading of work.

and bottom, one having an included angle of 120 degrees, the other, 90 degrees. These two angles will accommodate round or four-, six-, and eight-sided shapes. Member *E* has a series of threaded holes which are used to adjust the height of a clamp *F* to suit the bar stock being machined.

Clamp *F* is made of two pieces cut from flat stock and is held together with two rivets. Two spacers slightly over the thickness of clamp jaw *G* are placed between the pieces at the points where the rivets are applied. The clamp jaw has V-grooves to match those in the work-holders and a series of drilled holes that are used for height adjustment to accommodate a variety of stock sizes. The attachment is opened to receive the workpiece *H* by loosening screw *J*, swinging it outward and raising the clamp. Then, clamp jaw *G* is adjusted in height and screw *K* is tightened in place. The vertical position of the workpiece is adjusted by loosening the cap-screws retaining work-holders *D* and *E* and sliding these members up or down as required. For convenience, square-head screws having the same size as the one for the toolpost are used throughout.

Applications of the device include facing of workpieces with either a fly cutter or an end-mill mounted in the lathe spindle. Drilling and center drilling are also possible. A number of holes can be accurately spaced in a line by employing the micrometer dial on the lathe cross-slide. In addition, by swinging the compound rest, angular faces can be machined on the workpiece.

With a fly cutter and a center drill used in combination, workpieces can be rapidly faced and center-drilled on one end. Then, after resetting the carriage stop, the same operations can be performed on the opposite end, all pieces being uniformly produced. A stop of the type shown in

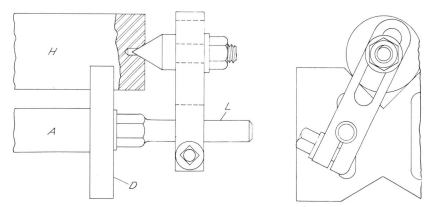

Fig. 22. Stop for locating workpieces that permits uniform center-drilling. Center may be replaced by a stud for positioning workpiece from face instead of center hole.

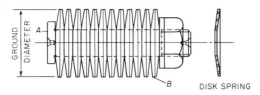

Fig. 23.   Expanding mandrel by mounting disc springs on an arbor.

Fig. 22 may be used for positioning workpieces. Special stud *L* replaces one of the screws holding part *D* to the base. Alignment of the device can be facilitated by first indicating the workpiece in a chuck and then setting up the attachment around the part.

Expanding Mandrel Made Up of Disc Springs

A novel expanding mandrel comprised of a series of disc springs mounted on a shoulder arbor is shown in Fig. 23. A nut and washer located on the right end of arbor *A* are applied to expand or contract the series of disc springs *B*.

In constructing the mandrel, the disc springs are slightly compressed. Then their outside diameters are accurately ground with the assembly held between centers. This type of mandrel has a strong holding power.

CHAPTER 5

Chucking and Workholding Devices for Lathes

Compound Rest Adjusts Lathe Tool to Diameters Dimensioned in Metric System

When the compound rest of a lathe is set at an angle of 23 degrees 11 minutes, each 0.001 inch of movement of the compound rest along its axis produces 0.01 millimeter of movement perpendicular to the lathe center line. This principle can be used to adjust a tool to a diameter dimensioned in the metric system.

Proof of the angular adjustment of the compound rest is explained by the diagram, Fig. 1.

Since 0.01 millimeter = 0.0003937 inch,

$$\sin \alpha = \frac{0.0003937}{0.001} = 0.3937$$

therefore

$$\alpha = 23 \text{ degrees } 11 \text{ minutes}$$

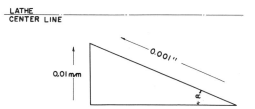

Fig. 1. When angle α = 23 degrees 11 minutes, a movement of 0.001 inch of the compound rest produces a movement of 0.01 millimeter (0.0003937 inch) relative to the lathe center line.

Fig. 2. A 1-inch gage-block has been placed against one sine-bar plug (for clearance), while the stack of gage-blocks against the other plug equals 4.937 inches, to obtain a difference of 3.937 inches.

A sine bar can be used to adjust the compound rest accurately, as in Fig. 2. After the right side of the compound rest is machined parallel to its dovetail slide, a plate is fastened to it. A 10-inch sine bar is then clamped to the plate. Using a cylindrical test bar between the lathe centers, the compound rest is swiveled until it is at the required angle of 23 degrees 11 minutes. This is obtained by placing gage-blocks between the sine-bar plugs and the test bar. The center distance of the plugs is 10 inches; so one plug has to be 3.937 inches farther from the test bar than the other plug.

Cam-turning Attachment

The production of cams in quantity can be a time-consuming and expensive operation, requiring much careful machining by skilled mechanics. With the aid of the attachment illustrated, however, cams may be readily made in a lathe from a prepared master. After the initial setup, the machining operations required to mass-produce the cams are routine.

Two equal-size sprockets, A and B, are mounted as shown in Figs. 3 and 4 — part A, on a true machined surface of the chuck and part B, on a shaft C. This shaft is supported in a bearing D secured by screws

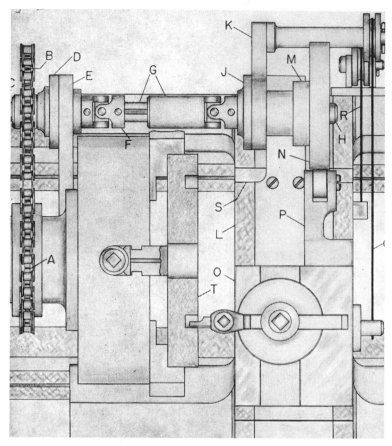

Fig. 3. Setup for turning cams on a lathe. Tool is guided by a master having the desired center.

to a hanger *E* that is clamped to the bed of the lathe. Motion is transmitted by a roller type chain. Universal joints *F* and a telescoping, intermediate coupling shaft *G* serve to drive shaft *H* while permitting both longitudinal movement of the carriage and lateral movement of the cross-slide. This arrangement allows shaft *H* to always rotate at a speed equal to that of the lathe spindle and in the same direction.

Shaft *H* is supported by a bearing *J* in a hanger *K* fastened to a special slide *L*. A collar *M* is keyed to the shaft *H*, and master cam *N* is attached to this collar with screws (Figs. 5 and 6). Slide *L* is fastened to the cross-slide in place of the compound. The screws that normally hold the compound are used for this purpose. Tool-block O is machined to fit slide L and has a T-slot for the toolpost identical in size to that in the compound. However, this slot is made parallel to the cross-slide for easier positioning of the tool. When slide L and block O are fitted together and mounted on the cross-slide, the top of block O is at the same height as the top of the assembled compound.

Roller-follower assembly P is attached to the block O, as shown. A weight secured to the end of cable Q is used to keep a constant and even contact between the roller and the master cam. A second weight attached to the cross-slide by cable R takes up any play between the cross-slide screw and nut. A block S which is machined to fit the dovetail

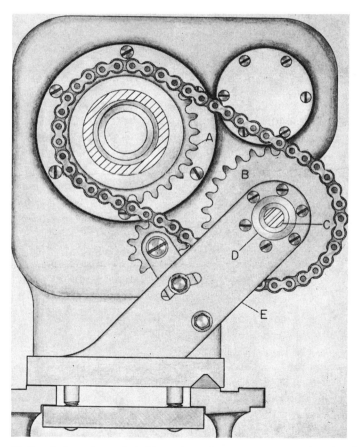

Fig. 4.   Roller chain and sprocket drive rotates the master cam at a 1 to 1 ratio with the lathe spindle.

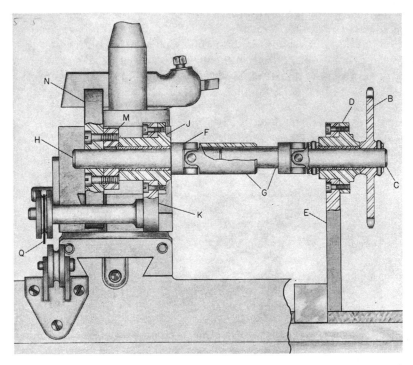

Fig. 5. Cam-turning device as seen from rear of lathe. For clarity, portions of the lathe have been eliminated.

slide of the saddle is fastened to and supports the overhanging end of slide *L*, as illustrated in Fig. 6.

Since the two sprockets are of equal diameter and pitch and are coupled by a roller chain, one turn of the lathe spindle will result in one turn of the master cam. When this cam is rotated, a tool mounted on block *O* may be used to reproduce the shape of the master in the workpiece *T*. Although the cross-slide and the carriage can be moved in the usual manner, the travel of the carriage is limited to the telescoping length of the shaft and sleeve. In addition to cams, this attachment may be used to copy other parts having irregular shapes. Of course, there are limits to the contours that can be reproduced in this way.

Radius-Bar Attachment for Lathe Operations

Machining concave surfaces requiring a high degree of accuracy is a difficult project for many small shops. The use of templates with the cut-and-try method is time-consuming, and in many cases the results are not too good.

The radius-bar attachment illustrated in Fig. 7 was designed for machining concave surfaces in a workpiece that is chucked in the lathe. With this device the maximum radius that can be produced is approximately equal to the maximum distance between the head and tailstock centers. Since the actual machining is done with the tool being moved across the workpiece by the power cross-feed, a high degree of accuracy and a good finish can be obtained.

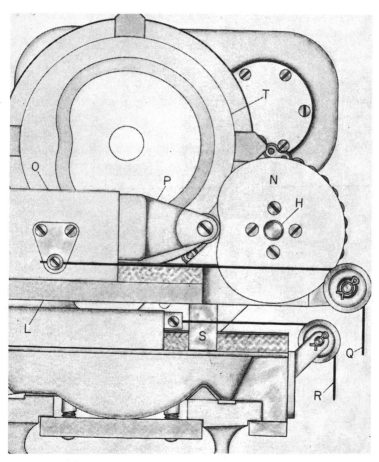

Fig. 6.   A weight is attached to cable Q to keep the roller follower constantly in contact with the master cam N.

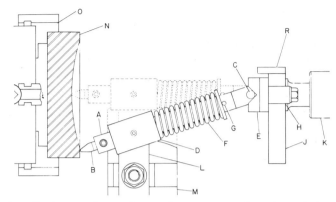

Fig. 7. A radius-bar attachment for turning concave surfaces in workpieces chucked in a lathe.

The radius of the finished surface is determined by the length of the bar A. Measurements for setting the bar are made with a height gage, the distance from the end of the tool bit B to the center of the pin C being made equal to the required radius. The bar is machined to slide freely through a bored hole in the sleeve D, and a tongue is milled on one end to fit a slot machined in the V-block E. As this slot is horizontal and perpendicular to the V-groove, any vertical movement of the bar is prevented.

A compression spring F and a washer are located between part D and a dowel pin G, which passes through the radius-bar. Spring F causes the pin C to be held tightly against the V-block and permits a sliding action between the bar and the sleeve when a cut is taken.

The V-block is attached by two clamps H to block J, which has a tapered shank that fits the bore of the lathe tailstock K. A keyway is milled in block J and a matching key is machined on the V-block. The key and keyway allow for adjustment of the V-block to off-center positions (Fig. 8).

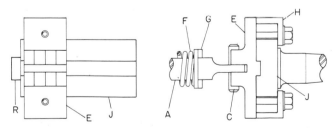

Fig. 8. Details of tailstock support for the pivoting arrangement of the radius-bar.

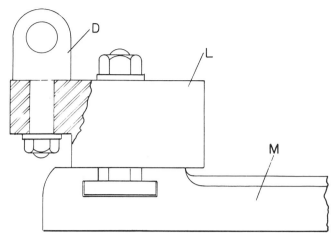

Fig. 9.   Power cross-feed drives the radius-bar by means of the pivoting sleeve *D*.

A nut, a washer, and the partially threaded shank on sleeve *D* are used to attach it to block *L* as shown in Fig. 9. The fit between these parts is such that the sleeve is allowed to swivel freely. Block *L* is bolted to the lathe compound *M* as indicated.

When machining a concave surface, the workpiece *N* is held in the chuck *O* and, starting at the outside of the workpiece, the tool is fed to the center by means of the power cross-feed. The depth of cut is regulated by the handwheel on the tailstock. To increase the cut depth, the tailstock spindle is moved toward the headstock. The attachment should be set up with the sleeve *D* placed as close to the tool bit as possible in order to prevent chatter. All moving parts should be lubricated freely.

To accurately locate the pivot point of the radius-bar, micrometer readings are taken over block *R* and V-block *E*. When the V-block is located central with the axis of the lathe, a reading should be taken and recorded. At this location, the radius-bar will produce a concave surface having its center in line with the lathe axis. The radius-bar can be offset to cut a groove of circular arc cross-section. In this case, the setover required is added to the initial reading of the micrometer.

Round nose tool bits having a radius of 1/32 inch and a cutting edge that is located at the horizontal center of the bar work well. The hole provided for the tool bit should be made deep enough to allow for some adjustment when setting the bar to cut the desired radius. It should be noted, however, that except where a minor adjustment of the tool bit is sufficient, a different bar will be required for each radius to be cut.

Toolpost Fixture for Radius Turning

Turning rounded ends on workpieces, such as punches for upsetting dies, can be simplified by using the toolpost fixture illustrated in Fig. 10. With this fixture, the desired radius is pre-set: all that remains for the operator to do is swivel the cutting tool by means of a convenient handle screwed into the tool-block.

A T-shaped base *A* fits in the toolpost slot of compound rest *B*. Two fillister-head machine screws *C* pass through the base and engage anchor plate *D*, thus providing a means of locking the fixture in place.

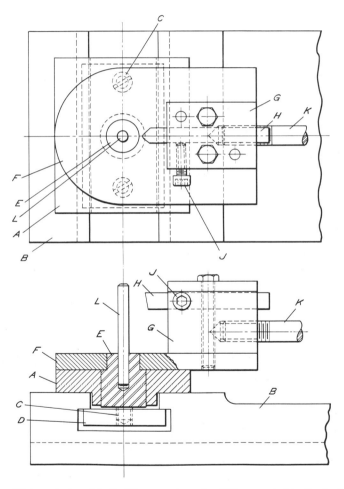

Fig. 10. Fixture designed to facilitate turning of radii is mounted in the toolpost slot of a conventional lathe.

A hole is drilled through the base to receive a stepped plug *E*. The larger diameter of the stepped plug is machined to provide a running fit.

The upper, or smaller, diameter of the plug is pressed into a hole drilled through rotating plate *F* so that both the plate and plug move as one unit. A tool-block *G* is bolted on the top surface of the rotating plate, as illustrated. It consists of a rectangular steel block with a groove milled across its upper surface to accommodate tool bit *H*. A plate is pinned on top of the grooved surface to form a permanent cover. Lock-screw *J* serves to hold the tool bit in place. For easy operation, a handle *K* is provided. It is threaded on one end and screwed into tool-block *G*.

To facilitate setting the cutting tool to the desired radius, pin *L* is employed. It fits in a hole located on the upper surface of plug *E*. The proper radius is measured between the center of the pin and the cutting edge of the tool bit. After the setting of this radius has been completed, and the cutting tool secured in place by means of lock-screw *J*, pin *L* is removed. The toolpost fixture is now ready for operation.

This fixture may be used with the compound rest set either perpendicular to the lathe bed or parallel to it. If perpendicular, the lathe carriage can be used to feed the cutter into the work; if parallel, feeding can be done with the compound rest. In either case, the radius is formed by manual rotation of plate *F* with the aid of handle *K*. By mounting the tool-block close to the center of the fixture so that the tool bit extends past the center, concave forms may be cut. With the tool block located in the position illustrated, convex forms may be cut.

Lathe Fixture for Eccentric Turning

A lathe turning operation was required to form two crankpins on a common center line, one on either end of a short shaft. Due to the small size of these cranks, it was not considered practical to turn them on centers. Also, the large main shaft diameter of the work prohibited the use of an eccentric collet on an engine lathe.

The fixture illustrated in Fig. 11 was designed to facilitate this turning operation. Body *A* is a weldment consisting of a rectangular block, a baseplate, and a supporting pad. The under side of the baseplate is machined to ride over a V-shaped bedway of the lathe, while the pad rides on a flat bedway. Four bolts and clamping plate *B* lock the fixture to the machine bed.

Cover plate *C* is hinged to the body and may be locked in place by means of latch-bolt *D*. A hole is bored through the body and cover combination to receive flanged bushing *E* with a running fit. Threaded collar *F* engages the right-hand end of the bushing and secures it in place.

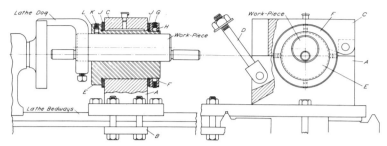

Fig. 11.   Lathe fixture simplifies turning of in-line crankpins located on both ends of a short shaft.

The collar, in turn, is locked by four set-screws *G* that force brass plugs *H* into contact with the bushing threads. A fiber thrust washer *J* is provided at each end of the body.

To use this fixture, the workpiece is placed within a hole bored through the flanged bushing at the desired offset and locked by four set-screws *K* and four soft plugs *L*. After placing the lathe dog on one end of the shaft, the bushing is inserted in the fixture body and the cover plate is closed and locked. Threaded collar *F* is run up snugly against the adjacent thrust washer and held firmly by set-screws *G*.

The right-hand crankpin may now be turned. When this is completed, the cover plate is unlatched, and the bushing reversed end for end. After reclamping, the second crankpin is turned.

Crank-Throw Fixtures for Offset Turning

Tooling for the turning of offset work surfaces, such as crankpin bearings, on a job-lot basis can be simplified by the use of adjustable crank-throw fixtures. One of these fixtures (used in pairs) is shown supporting the tailstock end of a crankshaft at X in Fig. 12.

Each of the two fixture units is constructed as shown at Y. Main frame *A* has a V-slot machined across its clamping face. Cap *B*, having a mating V-slot, is attached to the main frame by two bolts.

This clamping assembly is secured to a center-plate *C* by two studs *D*. The studs are welded flush with the reverse side of the center-plate and a light machining cut is taken across the welded face. A row of center holes *E*, view Z-Z, are then drilled on 1/2-inch centers along the machined surface.

Both studs *D* pass through a slot in the main frame. The slot is long enough to allow a 5/8-inch sliding movement of the main frame member when nuts *F* are loosened. This movement, together with the center-hole spacing on the bottom face of plate *C*, permits adjustments for a wide span of workpiece offsets. Although both fixture units are essentially the same, the head-stock member has a provision for driving from the faceplate.

Versatile Boring-Bar Holder for Lathe Operations

A holder that will accommodate various sizes of boring-bars without the necessity of using adapters, and which will automatically position the center of the bar in line with the lathe spindle, is shown at X in Fig. 13. The holder consists primarily of a steel block that has a V-groove on one side. The bottom of the vee is positioned at the same height as the center line of the lathe spindle.

Two clamps that are fitted with cap-screws are provided to hold the boring-bar in the V-groove. These clamps are made flat on one side to accommodate large-diameter bars, and have a high shoulder on the opposite side to permit adequate clamping of small-diameter boring-bars. A square washer is provided which fits the T-slot on the compound rest. This washer is mounted on a shouldered bushing that extends upward into the boring-bar holder. A spacer is ordinarily supplied between the compound rest and the holder, but this spacer may be omitted in cases where the height of the spindle center above the compound rest is a short dimension. A cap-screw held to the top of the block by means of a washer engages a tapped hole in the bushing to lock the holder assembly on the compound rest.

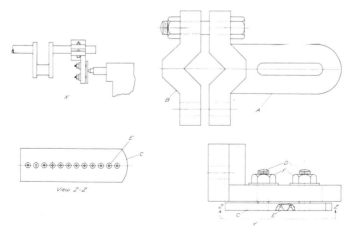

Fig. 12.   Offset turning on a small-lot basis is facilitated by using an adjustable fixture of this type at either end of work.

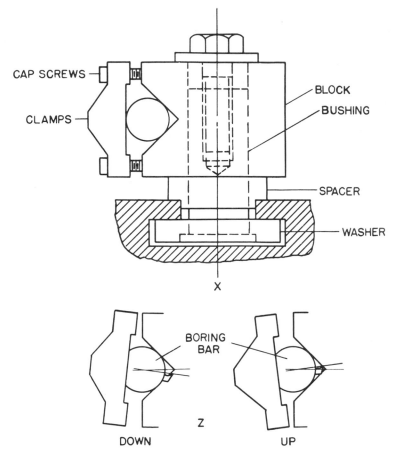

CAP SCREWS

CLAMPS

BLOCK

BUSHING

SPACER

WASHER

X

BORING
BAR

Z

DOWN   UP

Fig. 13.   Diagram showing the construction and operational possibilities of a versatile boring-bar holder.

An added advantage of this type of holder is that provisions can readily be made for locating the tip of the boring tool at the exact height of the spindle center line, or at desired amounts above or below the center line. To provide for such settings, a flat surface is milled on the bar exactly 180 degrees from the tip of the tool when it coincides with the lathe spindle center line. This will insure that the tool is automatically set to the desired height by simply placing the bar in the holder and tightening up the clamps.

Variations in the height of the tool tip can be obtained by tightening the lower clamp screws more than the upper screws, if it is desired that the tool point be high with respect to the spindle center line, as shown in the right-hand diagram at Z in Fig. 13. On the other hand, if it is desired that the tool point be low with respect to the spindle center line, the operator would tighten the upper clamp screws more than the lower clamp screws in order to obtain the condition indicated by the left-hand diagram at Z.

This boring-bar holder should be considered a permanent part of the accessories for any particular lathe because, due to the different heights from the top of the compound rest to the spindle center line on different machines, the unit is not generally interchangeable from one lathe to another. The conventional toolpost of the lathe is, of course, replaced by the boring-bar holder.

Faceplate Fixture Facilitates Precise Spacing of Bored Holes

A fixture that readily permits accurate spacing of holes to be drilled and bored in workpieces held on the faceplate of a lathe is shown in Fig. 14. The two inner perpendicular edges of an L-shaped frame and a special gage-block are used to locate the part after boring the initial hole.

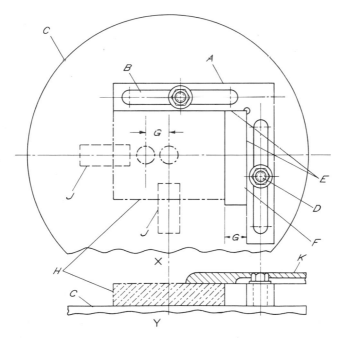

Fig. 14. Workpieces requiring holes having close-tolerance spacing can be quickly positioned for boring on a lathe face-plate equipped with this fixture.

The fixture frame *A* is made in one piece and ground all over. Slots *B* are cut in each leg of the frame to provide a means of securing it to the faceplate *C* with machine bolts *D*. The two inner surfaces *E* have to be precisely 90 degrees to each other and the base. A gage-block *F* is made and accurately ground to any desired spacing between the holes, see *G* (Fig. 14).

In use, the workpiece *H* is first secured to the faceplate with clamps *J*, shown dotted in view X, in position for boring one of the holes. The fixture and the proper gage-block are then bolted in place. The frame is carefully fitted against the workpiece as any misalignment between the part, the fixture frame, and the gage-block would cause inaccurate spacing. Accurate positioning can be accomplished by using cigarette papers as feeler gages between these three pieces. The fixture and the gage-block should be tapped in place with light blows of a soft hammer until the cigarette papers are snugly held between the surfaces in question. The special clamp *K*, shown in view Y, and the bolts are then firmly tightened, and the hole bored to size.

The second hole in the part is then accurately spaced simply by loosening the clamps holding the work, leaving the fixture frame firmly clamped, removing the gage-block and sliding the workpiece over into contact with the leg of the frame. This is the position left vacant by the gage-block spacer. The clamps are then tightened to hold the part and the hole is bored to size. A suitable counterweight is clamped to the faceplate of the lathe to balance the fixture and the workpiece.

Lathe Fixture for Machining In-Line Bores

Maintaining center-line accuracy between two in-line bores for anti-friction bearings without the costly machining of locating surfaces has always been a problem. Such an operation may be performed quickly and precisely with the indexing type of fixture illustrated in Fig. 15. This device was used originally in machining ball-bearing housings for mounting sawmill mandrels.

Faceplate *A* is the basic component of the fixture. Bolted to it is a ribbed cast bracket *B*, which is located positively by two taper pins *C*. The bracket is first machined on its mounting face adjacent to the faceplate. After bolting and pinning, the assembly is mounted on a lathe and the curved portion of the bracket, including a groove *D*, is machined.

A second cast bracket *E* has sides that taper from a wide, arc-shaped top to a clamping boss on its lower extremity. A tongue is machined

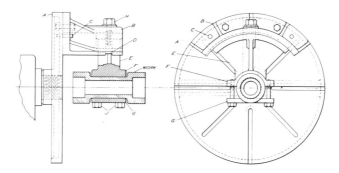

Fig. 15.   Lathe fixture with swiveling work-holder for boring in-line bearing seats.

along the arc to an accurate fit with its mating concave groove in bracket *B*. This is done by placing a stud between clamping jaws *F* and *G*, and holding it in a chuck.

Screw *H* passes through bracket *B* and is threaded into bracket *E*, thereby locking it in position. When this is done, a truing cut is taken on the internal clamping face of jaw *F* to bring the jaw to the diameter of the workpiece it is intended to hold.

In operation, the fixture is secured to the headstock spindle of a lathe. The workpiece is placed between clamping jaws *F* and *G*, and is held firmly in place by tightening four bolts *J*. In this manner, one bearing seat may be bored out and the end faced.

To bring the workpiece into position for line-boring the opposite end, screw *H* is loosened and bracket *E* rotated 180 degrees. Upon tightening the screw, the bracket will be drawn up snugly into the groove in bracket *B*, and the part will be centered with respect to the first bearing bore. It may be necessary to use counterweights for high-speed operation.

Two-Position Faceplate Fixture that Maintains Center Distance in Boring

Twin holes in gear housing castings were rough- and finish-bored on a turret lathe to an accurate center distance by means of the "rocker" type faceplate fixture shown in Fig. 16. The diamond-shaped fixture *A* is pinned to the faceplate *B* at *C*, around which it can swivel from the illustrated position, where the diamond lies to the left side of the plate, to an alternate position, where the diamond lies to the right side of the plate. A locating plug *D* in the fixture enters one of two bushings *E* in

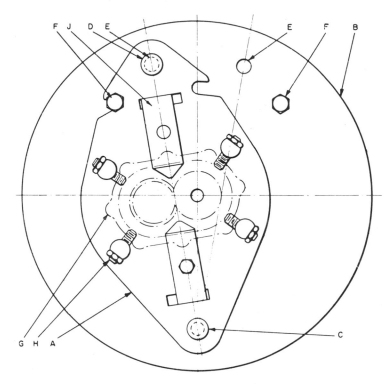

Fig. 16. Holes can be successively bored to an accurate center distance without resetting the work in the "rocker" type fixture.

the faceplate, after which a lock-screw *F* can be tightened to bind the
fixture to the plate.

Casting *G* is located in the center of the fixture, where it can be adjusted by set-screws *H* to bring one of the cores onto the center line of
the faceplate. After the work has been secured to the faceplate by
straps *J*, the first hole can be bored. To bore the second hole, the engaged
lock-screw is loosened, the locating plug is removed, and the fixture is
swiveled to its alternate position, where the locating plug can enter the
other bushing, after which the second lock-screw can be tightened.

The length of the arc between the bushings of the faceplate is accurately established, inasmuch as this will determine the precision of the
center distance of the holes. If in a similar application the operation is
to be performed at high speed, it would be advisable to fix a counterweight to the faceplate at an appropriate point, in order to compensate
for the off-center lay of the diamond-shaped fixture.

Chuck-mounted Bushing Facilitates Deep-Hole Drilling

When drilling deep holes in workpieces that are chucked in a turret lathe, the drill, if unsupported, may wobble and chatter. This condition can be eliminated or greatly reduced by means of a support bushing that is bracket-mounted to the lathe chuck. The arrangement illustrated in Fig. 17 is particularly designed to steady and guide long drills of the larger sizes.

Bracket *A* is of welded construction. A steel bottom plate is welded to a length of standard I-beam and another plate is welded to the top. After normalizing this weldment, a key is machined in the bottom to fit the T-slot in the chuck *B*. Two holes are then drilled for the T-bolts *C* which secure the bracket in place. A slot is milled in the top of the bracket to receive a swinging bushing plate *D* made of cold-rolled steel. In addition, the bushing plate is bored to receive a standard liner for a fixed renewable drill bushing *E*. Different drill sizes can, therefore, be accommodated simply by changing bushings. Lock screw *F* prevents the bushing from rotating.

A cross-hole is reamed through both the bushing plate and the bracket for a standard dowel *G*. This dowel acts as the hinge for the swinging bushing plate. Quarter-turn thumb-screw *H* is provided to hold the bushing plate in the operating position. For clarity, the workpiece is not shown in the right-hand view. When using a large-size drill *K*, the speed required may not be high enough to make counterbalancing of the chuck necessary.

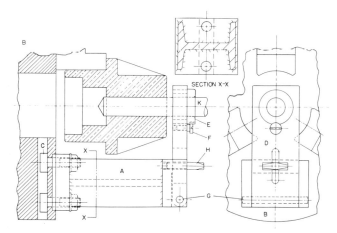

Fig. 17. Long drills can be guided and steadied by a bushing that is bracket-mounted to the lathe chuck.

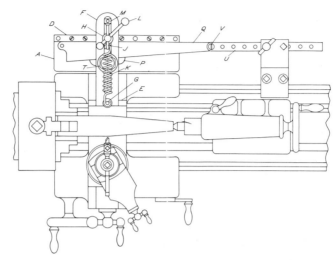

Fig. 18.   When the carriage feeds toward the spindle, follower *P* is forced back as it moves along the edge of template *Q*.

Spring-actuated Taper Attachment

Spring tension controls the action of a simple taper attachment that can be constructed for practically any lathe. By substituting a different follower, the device is easily adapted to use as a contour tracer. Plan and end views are shown in Figs. 18 and 19 (view W), respectively.

Platform *A*, of right-angle construction, is attached to the rear of the lathe carriage. The top surface of the platform is level with the top of dovetail *B* of cross-slide *C*. Guide *D* is doweled and screwed to the outer

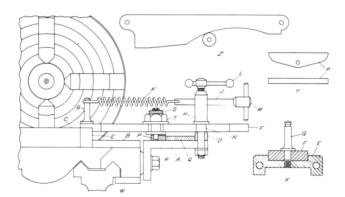

Fig. 19.   In end view W can be seen how the strength of spring *K* is adjusted through handle *M*. Views X and Y show the design of bridge-piece *E* and follower *P*, respectively; and view Z shows a typical follower and template for contour tracing.

edge of the platform, and is parallel to the lathe ways. The chip guard of the cross-slide was removed, and the cross-slide was released from its feed-screw nut.

Bridge-piece E (see Fig. 19, view X, for detail), fastened to the end of the cross-slide, has a channel on top which supports one end of link F. The other end of the link extends across the platform. The link is a snug fit in the bridge-piece, and is secured by spring-holder G. There is a slot through the link, made long enough to permit full travel of the cross-slide.

Post H, passing through the slot and contained in a hole reamed through the platform and guide, has a tension rod J hooked to the outer end of spring K. Adjustment of the spring is made by loosening clamp L and twisting handle M. Shoulder section N of the post is slightly longer than the height of the link and slightly smaller in diameter than the width of the slot.

Follower P (see Fig. 19, view Y, for detail) is kept snugly against template Q by the spring tension. The follower is engaged by threaded pin R, which in turn is screwed into bolt S fixed at a point in the slot by lock-nut T. Flats are milled on opposite sides of the bolt where it bears in the slot, so that it can be moved easily when the lock-nut is loosened.

The template is made from flat stock of a suitable thickness to clear the opening between the link and the platform. Taper of the template is one-half that of the work. For example, to produce a taper of 1/2 inch per foot, the template taper is 1/4 inch per foot.

At its right end, the template is anchored to an extension bar U which is clamped to the lathe bed. Linkage between the template and extension bar is through a shouldered screw V, which permits any needed swivel of the template for alignment. There is a hole tapped in each end of the template to receive this screw, so that the same template can be reversed and used to produce an opposite direction of taper in the work.

When setting up the job, lock-nut T is loosened and the cross-slide is positioned with the cutter close to the work. Then the lock-nut is tightened and final adjustment of the cutter made by advancing the compound rest. In operation, the follower moves along the edge of the template, which is stationary, and the cross-slide backs the tool away from the work as the carriage feeds toward the spindle. For tracing a contour, a roller type of follower and suitable template are substituted, as can be seen in Fig. 19, view Z.

Ball Prevents Damage in Taper Turning

When the tailstock center is set out of alignment with the headstock center for taper-turning on a lathe not equipped with a taper attachment,

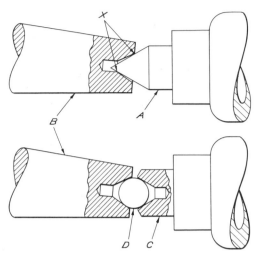

Fig. 20. When tailstock *A* is set off center, wear or breakage will occur at surfaces *X*. By placing a ball *D* between the work and a special center *C*, this condition is eliminated.

there is the possibility that the center-holes in the ends of the workpiece will be damaged. This is illustrated at the top of Fig. 20, where tailstock center *A* bears against surfaces *X* in the workpiece *B*.

This condition can be avoided by using a special tailstock center *C*, as seen at the bottom, and placing a steel ball *D* between the center and the center-hole in the workpiece. The same arrangement should be employed for the headstock center.

Internal and External Pipe Threads Cut Simultaneously on Lathe

Pipe threads have one outstanding characteristic in that the same pitch is used for a wide range of thread diameters. It is therefore possible to machine threads on different diameters in one operation — either internal or external threads, or both. Advantage is taken of this characteristic in machining the pipe fitting shown at X in Fig. 21. The fitting *A* is a brass casting with an external thread 4 inches in diameter and an internal thread 2 5/8 inches in diameter. On both sides of these diameters, 11 threads per inch are required to be cut.

The unmachined, large-diameter flange is gripped in a set of special shallow-step chuck jaws on a turret lathe. The machine happens to be equipped with a lead-screw of the same pitch as the thread to be cut. Two high-speed steel circular chasers with integral plain shanks are used as threading tools. Chaser *B* works with its cutting edge upside down at the back of the bore while the front chaser *C* works in the normal manner on the outside thread diameter. The chasers are so positioned in the holder *D* that both function at the same time. Two passes are sufficient to finish both threads to the desired diameters.

Prior to threading, the component is bored, turned, and chamfered by means of the tools mounted on one face of the hexagon turret *E*, as shown at Y. The tools are attached to a steel plate *F* which is bolted to the turret face. A boss welded to the plate is bored to receive bar *G* which carries a boring tool and a chamfering tool at its outer end. A second tool bar *H* carries an outside turning and chamfering tool. All of the turret tools cut simultaneously.

The turret tools are followed by a pair of facing tools mounted on the rear toolpost *J*, as shown in diagram X. Tool *K* faces the part on the front while tool *L* faces the flange and undercuts the thread end. All of the turning tools are carbide-tipped and operated at a surface speed of 500 feet per minute. The speed is reduced for threading, mainly to give the operator sufficient time to stop the operation before the chaser runs

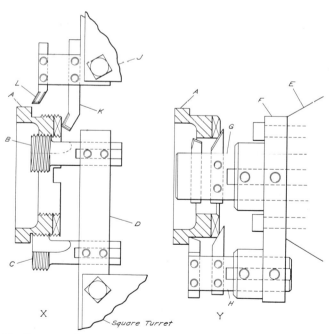

Fig. 21. Tooling setup for cutting internal and external pipe threads at the same time on a turret lathe, view X; gang of tools set up for turning, boring and chamfering the pipe fitting before threading, view Y.

up against the shoulder. The floor-to-floor time for each part is three minutes.

Multiple Thread-cutting Attachment for Lathes

In cutting threads having more than one lead several different methods are applicable for correctly locating the second, third, or fourth thread on the work. In many cases the work is rotated while the spindle and lead-screw remain fixed. Sometimes the work and spindle together are revolved the required 1/2 or 1/4 turn, while the lead-screw remains fixed. The latter method is used for internal threads.

With the attachment shown in Fig. 22 nothing is moved or changed except tool-bar A in cutting internal or external multiple threads. The movement of the tool-bar is accurately measured by means of a micrometer to position the cutter for each consecutive thread lead. In Fig. 22, one thread has been completed on work B with the lathe set up for cutting three threads per inch at a pitch of 0.3333 inch. A second thread will be cut between the lands of the present thread to produce a screw with six threads per inch having a pitch of 0.1667 inch.

When the first thread has been finished, a reading is taken on the feed-screw dial of the compound rest, and a micrometer reading is taken over parts C and D. Part C is a collar integral with a small-diameter shaft which is fixed in block F, while part D is a collar which is adjustable along the tool-bar.

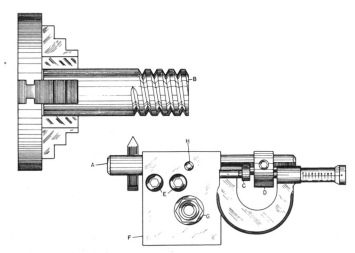

Fig. 22. Attachment designed to facilitate the cutting of multiple threads on a lathe.

Two socket-head screws *E* are then loosened to permit the tool-bar to be moved 0.1667 inch either to the right or left to correspond to the difference in pitch between the two thread leads to be produced. This movement is checked with the micrometer. The screws are then tightened and the second thread is cut. The amount of movement that is imparted to the tool-bar for any job is determined by dividing 1.0000 inch by the number of threads per inch. For example, in cutting a four-lead thread having 16 threads per inch, the amount of tool-bar movement in positioning for starting the second, third, or fourth threads would be determined by dividing 1.0000 by 16, or 0.0625 inch.

Block *F* is fastened to the compound rest by means of a stud and nut, indicated at *G*, with the tool-bar being mounted parallel to the axis of the lathe. The block is bored to a sliding fit to receive the tool-bar. A slot is milled in block *F* to permit clamping the boring-bar by tightening screws *E*. Set-screw *H* engages a keyway in the tool-bar to prevent the latter from turning during an operation.

For cutting internal threads in a part, the tool bit would be placed in the bar with the cutting edge on the side opposite to that shown in Fig. 22.

Lathe Tailstock Adapter for Solid Taps and Hand Reamers

Tapping and reaming operations are frequently performed in a lathe. Where a solid tap or hand reamer is used, it can be aligned and supported on the spindle center line by the bearing of the tailstock dead center in the conical opening in the shank end of the cutter. Ordinarily an open-end wrench is held on the squared end of the shank to prevent rotation of the cutter. This practice is satisfactory, although somewhat awkward, in that it requires the use of both hands of the operator — one to maintain the wrench in position and the other to advance the tailstock. Then too, a hard spot in the work hole might bind the cutter, jamming the operator's fingers between the wrench and the lathe bed.

Both of these undesirable circumstances can be avoided by means of the simple adapter for the tailstock illustrated in Fig. 23. It consists of a collar *A* and a U-shaped yoke *B*. The collar fits the tailstock spindle *C*, being feathered to the keyway *D* and fixed against longitudinal movement by a thumb-screw *E*. The legs *F* of the yoke fit slots in opposite sides of the collar, and in which they are secured by screws *G*. A squared hole *H* in the central section of the yoke receives the squared end of the cutter shank.

The hole *H* is made large enough to accommodate a range of tap sizes. The tap is aligned and supported by the dead center *J* in the customary

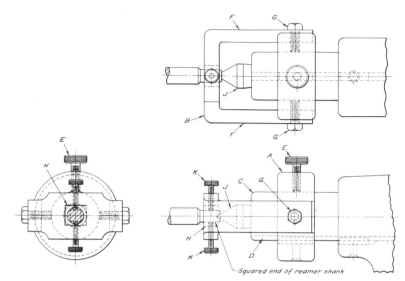

Fig. 23.   This adapter dispenses with the awkward, often dangerous, practice of using a wrench to prevent a solid tap or hand reamer from rotating.

manner. Set-screws K are adjusted radially to contact opposite sides of the shank square, thereby preventing rotation of the taps.

Lathe Tailstock Adapters for Threading Small Parts

Threading of small parts on a lathe can readily be accomplished by means of the two tailstock adapters illustrated in Fig. 24. These handy attachments prevent forcing the lead or stripping the threads, and are especially suitable for fine threading.

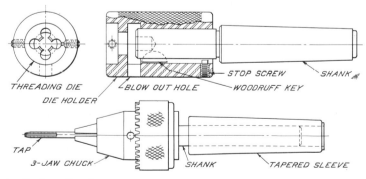

Fig. 24.   Lathe tailstock adapters for external threading (above) and for tapping or drilling (below). The die or tap can slide to prevent stripping the thread.

The adapter for external threading, seen at the top, consists of a die-holder and a hardened and ground shank that fits into the lathe tailstock. A keyway in the bore of the die-holder permits the holder to slide along a Woodruff key on the shank. At one end of the die-holder there is a set-screw that acts as a stop and prevents the holder from sliding off the shank. Accumulated chips can be blown out by compressed air directed through a hole provided in the bottom of the holder.

The tap or drill adapter, shown at the bottom, consists of a chuck having a hardened and ground shank which slides within a tapered sleeve. The sleeve fits into the lathe tailstock. Rotation of the tap is prevented by gripping the knurled periphery of the chuck.

Rethreading Attachment for Removing Burrs after Slotting Threaded Work

Figure 25 shows an attachment designed for rethreading work in a lathe to remove burrs after a slotting operation. The attachment fits the cross-slide of the lathe, and slides in dovetailed gibs to carry the die-holder to the work.

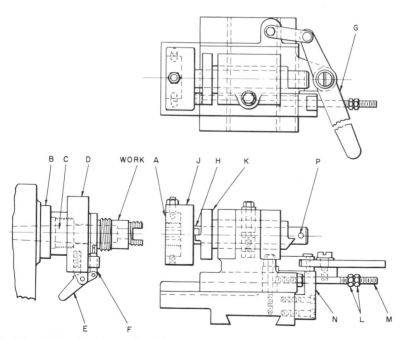

Fig. 25. Lathe attachment simplifies a rethreading operation performed to remove burrs from work that is slotted after being threaded.

One of the principal requirements in this design was that the threading die *A* have a sufficient number of lands to insure adequate support on the unslotted portion of the work as it revolves. For this purpose, the die is provided with eight lands. The lathe spindle, centering arbor, and work-holding fixture are designated *B*, *C*, and *D*, respectively, in the illustration.

In operation, the lever *E* is depressed to withdraw a locating and holding pin *F*, so that the work can be placed against the shoulder of the fixture. Then a hole in the work is aligned with this pin, which enters it to secure the work on the fixture. A handle *G* on the lathe attachment is now actuated to move the die into contact with the revolving work. As soon as the die engages the threaded portion of the work, it is automatically fed forward until the driving lugs *H* on the die-holder *J* disengage from the lugs on an adjustable reversing sleeve *K*. At this point in the cycle, the die will, of course, revolve.

The disengagement of the driving lugs is accomplished by contact of adjustable stop-nuts *L* on stop-rod *M* with a pad *N* as the attachment moves forward. The location of these nuts on the rod regulates the point at which the die is released. When this point has been reached, a shoulder on the shank of the reversing sleeve *K* engages a pin *P* in the die-holder shank, thus disengaging the die from the work. At the same time, the lathe spindle is reversed. The lathe attachment is moved back to its original position by actuating handle *G*.

Jaw Thread Sections Formed on Lathe

The back of each jaw of an independent chuck has the form of a nut section which engages a threaded spindle. Turning the spindle adjusts the jaw radially in the chuck body. It is possible to obtain the thread form of the jaw in a lathe operation, as shown in Fig. 26.

Chuck jaw *A* is supported over compound rest *B* and is raised to the machine center line by filler block *C* and baseplate *D*. The baseplate is drilled and counterbored to retain cap-screw *E* which engages toolpost T-nut *F*. To hold the jaw in place, a tapped hole in the baseplate receives stud *G* running through clamp *H*.

The work surface of the jaw is indicated and set parallel to the machine spindle. This is done by swiveling the compound rest in the required direction.

Machining is performed with boring-bar *J* held in spindle collet *K*. The bar must be at least twice the length of the work surface and is supported at its outer end from tailstock center *L*. First the surface is made concave, with a radius corresponding to the minor diameter of the

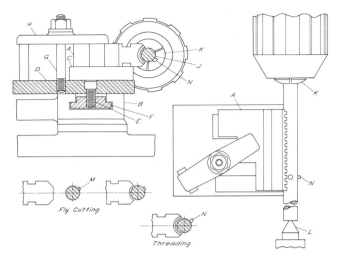

Fig. 26.   The back jaw form is obtained by fly cutting and single-point threading with cutter held in a boring-bar and driven by the lathe spindle.

thread, by fly cutter *M*. Then this cutter is removed, and the thread is single-pointed with form cutter *N*.

For the fly cutting, the carriage longitudinal feed is used. Correct radius is obtained by extending the cutter an amount equal to the thread depth. This can be checked with a micrometer. For example, if the bar is 0.625 inch in diameter and the minor diameter is 0.875 inch, the micrometer reading across the bar to the cutter tip should be 0.75 inch. The thread illustrated is square, so the form cutter is ground to the correct pitch, and the lead-screw of the lathe is used for the single-pointing.

V-Block Lathe Follower-Rest

A homemade lathe follower-rest consists of a simple support for a V-block that takes the cutting thrust. In addition to the V-block, a few pieces of scrap-plate are required. The device is attached directly to the compound rest of the machine.

In Fig. 27, baseplate *A* is bored to a loose fit over the tool-post, being secured in position by the toolpost assembly. V-block bracket *B* is welded to the baseplate, and its upper part is bent approximately 45 degrees, as illustrated. Back-stop *C*, located a short distance behind the bracket, is similarly welded to the baseplate and bent. Added rigidity is furnished by gusset *D*, welded to both the baseplate and the back-stop but without physical connection to the bracket.

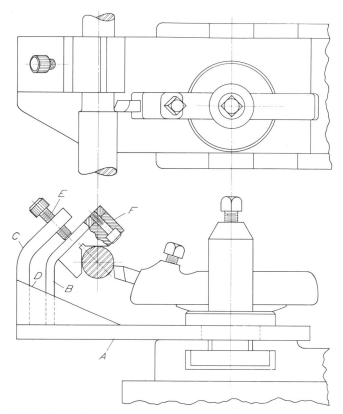

Fig. 27.   Inclined at 45 degrees, the V-block prevents the work from climbing over the cutter.

A tapped hole in the back-stop contains pressure-screw *E* which bears against the bracket and provides for fine adjustment of V-block *F* secured to the front of the bracket. The V-block may be either steel or bronze, depending on the nature of the job.

To set the follower-rest, the cutter is loosened in the tool-holder, and the cross-slide of the lathe is run up until the V-block makes contact with the work. The cutter is adjusted by tapping it forward by light hammer blows and tightened in the tool-holder. Then, pressure-screw *E* is advanced or retracted for fine adjustment. For exact calibration of cutter infeed, a dial indicator can be clamped to the baseplate and brought into contact with the rear of the tool-holder to record the amount the tool-holder is advanced.

The purpose of having the body of the V-block at an incline is to prevent the work from climbing over the cutter. For extremely fine

Fig. 28.   Makeup of the phantom center is shown with the universal chuck on the left and the adapter with its mating thread on the right. The ground shank for steadyrest use and the 60-degree center seat show at the extreme right on the adapter.

finishing with light cuts, a V-block of hard maple, end-grain contacting is recommended. It should be well lubricated in use.

Phantom Center Permits Turning Big Tubes in Engine Lathe

A small universal chuck and an adapter, Fig. 28, combine to make a tailstock bearing that extends the applications of engine lathes. Screwed into the chuck back-plate, Fig. 29, the arrangement permits turning large tubes on an engine lathe, using the 60-degree center seat (Fig. 28) as a tailstock bearing. The headstock end of the tube is held and driven by a three-jaw chuck, as in Fig. 29, in typical fashion for turning the outside diameter of a cylinder.

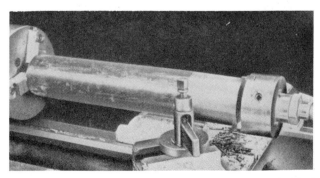

Fig. 29.   Phantom center in use, turning a length of tubing with the 60-degree center seat of the adapter supporting the work and tail chuck in conventional style.

Another application for the adapter is when a tubing workpiece is too long to be held between the regular centers. Used in this way the chuck and its adapter are supported by a standard steadyrest. The steadyrest is seated at the extreme end of the lathe bed, Fig. 30, with the tailstock removed. The shank surface of the adapter is ground concentric with the axis so that it makes a good bearing.

Adjustable Center Designed for Use in Headstock Spindle of Turret Lathes

In turret lathe operations it is sometimes desirable to support one end of the work on a center mounted in the headstock spindle, and to take the cuts on the part as close to the spindle as possible so as to insure maximum rigidity and size control. In cases where the cuts are taken at a considerable distance from the end of the work, some method of extending this work end into the spindle and supporting it on an internal center would greatly facilitate the operation.

Figure 31 shows a simply constructed adjustable type center unit designed specifically for this purpose. It will handle a large variety of work and hold parts with extreme accuracy. The unit consists primarily of a cylindrical steel plug A which is machined to a slip fit in the spindle. One end of the plug is reamed to provide a press fit for the shank of a hardened center B. The other end is drilled to receive the ball-socket end of screw C. This screw can be made to any required length and is sup-

Fig. 30.   Workpiece, too long for tailstock centering, is held for turning by the steady-rest mounted at the end of the lathe bed and bearing on the adapter shank.

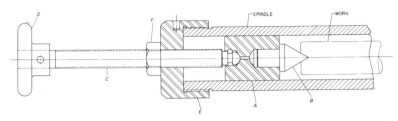

Fig. 31.   Adjustable center unit designed for holding work inserted into the end of the headstock spindle on turret lathes when cuts are to be taken close to the spindle.

plied with a handwheel *D* at the outer end. The screw engages a tapped hole in round cap *E* which serves to attach the unit to the outer end of the headstock spindle. By turning the handwheel, plug *A* with center *B* may be adjusted longitudinally within the spindle to suit the work. An internal thread in cap *E* fits the rear end of the headstock spindle.

In attaching this adjustable center unit to a machine, the rear cap of the headstock spindle is first removed; then the center unit is slid into the spindle and cap *E* tightened against the end of the spindle. After the end of the work is entered into the headstock spindle in loading work into the turret lathe, the adjusting screw is advanced to position center *B* for supporting the work firmly. Lock-nut *F* is then tightened against cap *E* to hold the unit securely in place.

Lathe Tool Bits Held Vertically in a Modified Holder

A modified form of lathe tool-holder in which standard square or rectangular tool bits are held in practically a vertical plane is illustrated in Fig. 32. Cutting is done by the upper end face of the bit, which is ground to provide the desired rake angles and round nose. With this design, the main cutting loads will be directed substantially along the full length of the bit, thus insuring a high degree of stiffness and resistance to deflection. Such a holder permits heavy stock removal, high rates of feed, and a good surface finish, with minimum tool breakage and long tool life between sharpenings.

Shank *A*, which is an integral part of tool-holder *B*, is machined to fit the toolpost of a lathe. A shoulder *C*, between the body and the shank of the holder, can be set to bear against the front of the toolpost body. The angular slot *D*, machined in the face of the holder, provides a sliding fit for the tool bit *X*. The compound angle of the slot (see upper view) provides the required end clearance (7 1/2 degrees) and side relief (15 degrees) for the tool bit, and it is only necessary to grind the back-rake

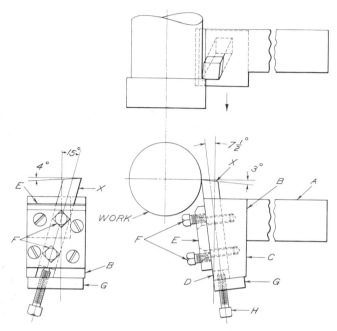

Fig. 32. Lathe tool-holder in which tool bit X is held in practically a vertical plane. With this arrangement, heavy stock removal is possible at high rates of feed.

angle (3 degrees) and side-rake angle (4 degrees) on the top face of the bit. These angles or the angular disposition of the slot can be varied to suit the material being cut and other conditions.

The tool bit is confined within the slot by plate E, which is secured to the front face of the holder by four screws and held in the desired position by two clamping screws F that bear directly on the bit. Permanently secured to the bottom of the holder by screws or by welding is a stop-plate G. Threaded through this plate and bearing on the bottom face of the tool bit is a long set-screw H, which is used to adjust the height of the bit in the holder. When the bit becomes short, it can be raised in the holder by inserting back-up blanks between the bit and the set-screw.

Tandem Cutoff Tool

Workpieces to be form-turned on a turret lathe can often be produced more economically if one form tool is used to produce two or more components. If the design of the part is compatible, a special tool-block such as the one illustrated at *A* (Fig. 33) can be made to hold two cutoff tools. The work can be severed from the bar as well as from the adjacent piece in one radial feeding movement.

The tool is designed so that the right-hand cutter leads the left-hand one by a slight but sufficient amount. The tool cuts off the component shown at the right just before the cutter at the left parts the remaining piece from the bar. A kick-off spring *B* placed between the cutters pushes the left-hand piece out from between the tools at completion of the dual operation.

Lever Operates Two Undercutting Tools

Two quick-acting tool-slides speed a lathe operation of undercutting both sides of a shoulder preparatory to grinding. In Fig. 34 the workpiece *A*, already faced and bored, fits over arbor *B*. A key *C* engages a keyway to drive the workpiece.

Slides *D*, containing undercutting tools *E*, are mounted in blocks *F* on the lathe carriage. These blocks are toed in toward the shoulder. Each

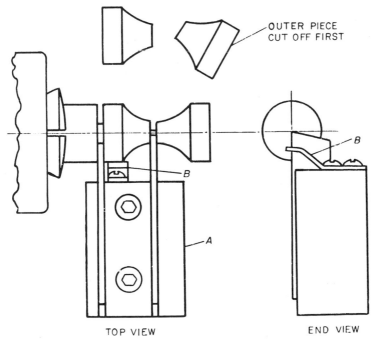

Fig. 33.   Turret lathe for parting two form-turned workpieces in a single operation.

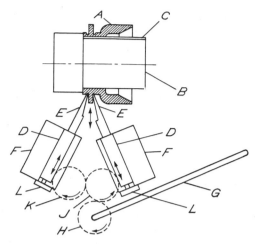

Fig. 34.   Pulling lever *G* causes slides *D* to move in, and tools *E* undercut the shoulder

slide is equipped with a gear rack. Lever *G* controls gear *H*, which drives gear *J*, in mesh with the right-hand slide rack. Gear *J*, in turn, drives gear *K*, in mesh with the left-hand slide rack.

Thus, pulling the lever moves both slides in, and the tools undercut the shoulder. Depth of cut is adjusted by stops *L* on the slides.

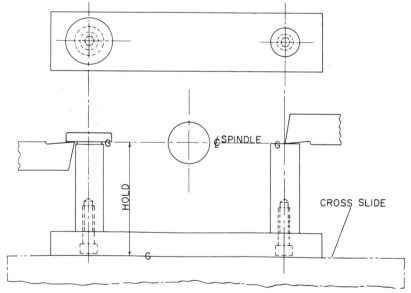

Fig. 35.   Simple gaging device for checking the height of front and rear cross-slide tools.

Gage for Setting Cross-Slide Tools

In using cutting tools at both the front and rear of a lathe cross-slide, it is important to make certain that the cutters are exactly in line with the center of work if maximum efficiency is to be obtained. Figure 35 shows a simply constructed gage that will greatly speed the setting of such tools and eliminate time-consuming trial-and-error methods.

Two round hardened plugs are secured to a base by means of socket-head cap-screws. The plug for the front cutting tool is made with a shoulder at the top end. The distance from the bottom surface of the head on this plug to the bottom of the base is made identical to the dimension from the center line of the spindle to the top of the cross-slide. Tools at the rear of a cross-slide must be used upside down, and so the other plug can be made without a shoulder.

Lead-Screw Dial Locates Lathe Carriage

To advance the carriage of a small engine lathe along its bed in precision movements, the end of the lead-screw was fitted with an accurately graduated dial and a knurled knob. In use, the lead-screw is disengaged from the change-gear train, and the half-nuts in the carriage apron are closed around the lead-screw. Then the dial on the end of the lead-screw is rotated manually by means of the knurled knob.

Fig. 36.   To accommodate the dial, last few threads of lead-screw are machined down and support bracket is moved to the left.

The dial comes in handy for jobs like spacing a series of grooves or shoulders along a shaft or boring a hole to an exact depth. In addition, the dial can be used to read carriage movements while it is in slow power feed.

To adapt the lead-screw, it was removed from the lathe, and the threads near the end (seldom used anyway) were machined down to the end journal size for a distance of about 1 1/4 inches. The support bracket was then moved to the left the same distance and resecured to the bed, (see Fig. 36).

The dial has an integral stem which is bored to fit the end of the lead-screw. A slot along the stem wall and two set-screws provide the clamping means. A pointer is attached to the bottom of the front way.

At the start of a feeding movement, the set-screws are loosened and the dial is rotated independently to a zero setting with the pointer. Then the set-screws are tightened and the dial is bound to the lead-screw.

Novel Method for Producing Small Pins

A device that can be used in a bench lathe to manufacture small chamfered pins at a rapid rate is shown in Fig. 37. The unusual feature of this arrangement is that the tools rotate instead of the stock.

In construction, the device incorporates a tube *A* which is gripped by a collet or chuck in a small lathe. A tube *B* is made to press fit in tube *A*. The right-hand end of tube *B* rotates in a bushing in a block *C* bolted to the lathe compound. Handle *D*, positioned approximately vertical, pivots on a pin in a block *E* clamped to the lathe bed. This handle swings left and right, and pushes a sleeve *F* horizontally along tube *A* by means of a shouldered screw secured in the sleeve. This screw is retained and slides in a through slot in the handle.

Part *G* (a collar pressed on tube *A*) carries a projecting block *H* which is turned on the left-hand end for a press fit in a counterbored hole in

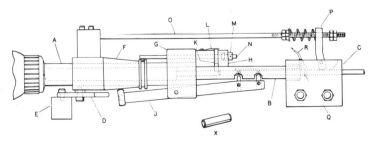

Fig. 37.   Plan view of device that is used for rapid production of small chamfered pins on a bench lathe.

the collar. In addition, collar G is slotted to accept an arm J that is pivoted on a dowel mounted across the slot. This arm supports a roller on the left-hand end and two tools are carried in slots in the right-hand end. The roller rides on sleeve F which controls the infeed motion of the tools. A V-notch is cut by the right-hand tool, and the second tool parts the stock at the previously machined notch. A completed pin is seen at X.

The stock-stop mechanism consists of a stop-plate K that slides in a slot in block H. Stop K is retained by a rod N and carries a pin which enters the tube to stop the workpiece. Spring L holds the stop-pin in the tube. Collar M is chamfered on the left-hand end to provide a means of retracting the stop. On the back of sleeve F is a rod O which operates stock clamp P.

In operation, the stock which does not rotate is manually fed in at the right, passing through block C and a neoprene ring Q, and coming to rest against stop K. Ring Q prevents the coolant from flowing back along the stock. Tubes A, B, and associated parts G, H, J, K, L, M, and N rotate with the lathe spindle. Then, handle D is pushed to the right, moving sleeve F to the right. This causes the roller on arm J to ride up on the taper and thus feed the tools into the work. The action of clamp P prevents the work from rotating as the tools simultaneously cut off one pin and chamfer the next.

Coolant flows through tubing R into block C and along a keyway in tube B, lubricating the bushing in which tube B rotates and flushing chips away from the two tools. The tool area is covered to prevent coolant from being thrown outward at the operator. This cover is not shown.

On completion of the cuts, handle D strikes a stop (not shown). Then, as the handle is returned to the left, the tools are retracted automatically by centrifugal force, and clamp P is released. Further movement of the handle to the left causes the tapered end of collar M to enter a hole in stop K and retract the stop. When the stock is again fed toward the headstock, the finished piece is pushed to the left along the inside of the tube. Then, by pulling the stock back slightly, and moving the handle a short distance to the right, the stop is allowed to re-enter the tube. After pushing the stock against the stop, the work cycle is repeated. Finished pieces fall out at the rear of the lathe spindle into a collecting pan.

Rotating Lathe Cutter-Head Designed to Machine Awkward Trunnions

A swing cutter-head designed to turn trunnions on a structural steel workpiece having an extension too long to be rotated on a machine tool

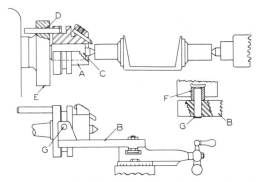

Fig. 38.   Rotary cutter-head which is moved longitudinally in a lathe operation.

is shown in Fig. 38. The trunnions are welded to opposite sides of a structural steel channel 6 feet long. Used on a medium-sized engine lathe, the special cutter-head proved economical for the amount of work machined.

After laying out and drilling center holes on an axial line common to each of the trunnions, the work is placed between the headstock and

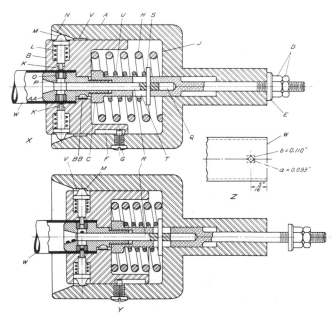

Fig. 39.   This turret-lathe tool punches two holes simultaneously near the end of a thin-walled tube.

tailstock centers of the lathe. The long channel arm is supported by a steel trestle set on the floor behind the lathe.

Rotation of the lathe spindle turns the cutter-head body A as it is carried along the work longitudinally by link B which connects the body with the tool-rest on the carriage. Body A is a snug sliding fit on arbor C, and the latter is provided with a tapered shank to fit the headstock spindle.

The cutter-head is driven by a pin D which is located in driving plate E. An annular groove around the periphery of the body serves as a track for roller F which rides on pin G. When the carriage feed is engaged, body A is traversed longitudinally along the work, and, at the same time, rotates to make the cut. The trunnions are machined for about one-half their length.

The correct setting of the tool bit for both the rough and finish cuts can be made by the use of a stepped gage bar held between centers. In chamfering the sharp corners on the ends of the trunnions, the hand feed is used.

The work is turned end-for-end to machine both trunnions. However, because the job is performed with the work held between centers, the axial alignment of both trunnions can be maintained readily.

Turret-Lathe Tool Pierces Thin-walled Tubes

A piercing tool designed for use in a turret lathe cleanly pierces two square holes of different size in the walls of a light-gage, annealed brass tube. Cross-sections of the tool before and after the tube has been pierced are shown in views X and Y of Fig. 39. Only low piercing pressures are required because of the softness of the material, and the tool design proves effective on tubes having a wall thickness of 1/32 inch. As shown in view Z, hole a lies diagonally to the axis of the tube, while larger hole b is parallel.

The two holes are pierced simultaneously by punches K. These punches are pressed radially inward by beveled surface N of the bore of the housing A as it is advanced by the turret over sleeve B. The sleeve end-wall at the left is counterbored 1 inch deep to a diameter 0.004 to 0.005 inch above the outside diameter of tube W. Plug AA, ground on its larger diameter from 0.002 to 0.003 inch less than the bore of the tube, is a light press fit within the sleeve, and is restrained from turning by a Woodruff key BB.

Holding the plug against axial movement is a hollow shaft C which is screwed over the threaded shank of the plug. At this end, the shaft has a hexagon head for locking purposes. On the diameter protruding from

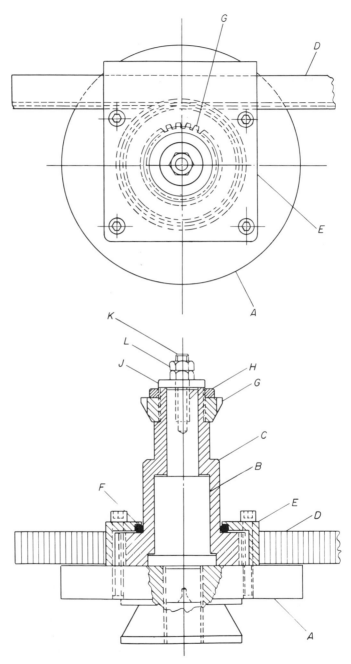

Fig. 40. Rack and pinion drive rotary cutting tool on this lathe attachment. Cutter is mounted on a special toolpost assembly that replaces the compound slide.

the housing shank, the shaft is threaded for lock-nuts D, and contains washer E.

The sleeve is free to slide within the bore of the housing, but is kept from rotating by set-screw G extending into slot F. A compression spring H keeps the sleeve in its maximum left-hand position when the tool is not operating.

At that time, the punches are raised by the action of springs L. The outer end of each punch has a hardened dome M which bears uniformly against beveled surface N. To insure clean and accurate piercing, bushings O conform closely to the shape of their respective punches and are a press fit in counterbored holes on opposite sides of the plug. Punch slugs are pushed into holes P.

To eject the slugs, a blind hole Q is provided in the shaft, and rod R is fitted to slide within the blind hole and the bore of plug AA. Steel pin S, fixed across the rod, moves along slot T in shaft C. Coil spring U holds the rod normally retracted to the right. Thus, the rod moves in unison with the sleeve and the shaft for a certain distance.

In operation, the piercing tool is mounted in a face of the hexagon turret, with the punches preferably in a horizontal plane, to facilitate the ejection of slugs. When the bottom of the hole in the sleeve contacts the end of the tube, further advance of the turret causes the housing to slide to the left over the sleeve, and the punches move simultaneously through the tube walls. Slugs are pushed into the bore of the shaft.

While the punch domes are constrained by diameter V, the inner wall of the housing bears against pin S, making the rod move to the left to sweep out the slugs, as in view Y. Upon retraction of the turret, the various springs return the elements of the tool to their initial positions.

Lathe Attachment Drives Rotary Cutter

The cutting action of a standard lathe involves the passing of a fixed tool along the surface of a rotating workpiece. There are instances, however, where rotary cutters are used to advantage for producing various shapes on a lathe.

A rack-and-pinion attachment that permits such a tooling setup to be made is illustrated in Figs. 40 and 41. This particular device was designed to generate cooling fins around the outer surface of a cylindrical housing. The arrangement includes a toolpost assembly, Fig. 40, that replaces the compound slide. Working in conjunction with the special toolpost is a support bracket, Fig. 41, that is clamped to the lathe bed. Also shown in Fig. 41 is the workpiece following machining.

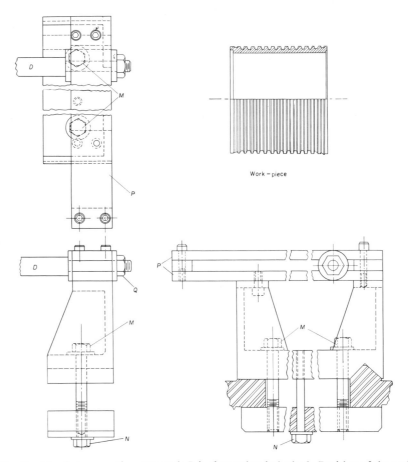

Work – piece

Fig. 41.   Bracket supporting gear rack *D* is clamped to lathe bed. Position of the rack can be adjusted by sliding it between guide plates *P*, then tightening nut *Q*.

Base *A* of the toolpost assembly (Fig. 40) is machined to permit mounting on the cross-slide of the lathe. Threaded into this base is post *B*, over which a sleeve *C* is accurately fitted. The flanged lower end of the sleeve is cut as a spur gear that meshes with teeth on a long rack *D*. Cover plate *E*, held to the base with machine screws, protects the teeth of both gear and rack from chips. An O-ring seal *F* increases this protection, while additional cover plates (not shown) shield exposed portions of the rack.

Cutter *G* threads onto the upper end of the sleeve and is held in place by lock-nut *H*. The sleeve, in turn, is held to the post by washer *J*, stud *K*, and adjusting nuts *L*.

The support bracket (Fig. 41) is clamped across the lathe bed by tightening bolts *M*. Bolt *N* merely serves to hold the bracket together when not in use. The right-hand end of the gear rack *D* is free to slide between two guide plates at *P*. Length of the guide plates is sufficient to permit gear-rack adjustment to accommodate various diameters of workpiece. Also, the rack is long enough to suit any length of workpiece within the capacity of the lathe.

In operation, nut *Q* is loosened to permit cross-feed adjustment of the tool, then tightened for traverse feed. With the rack thus held stationary, traverse movement of the lathe carriage imparts a clockwise rotation to the cutter as it shapes the rotating workpiece.

Although the illustrated device was intended for one particular application, there are numerous other operations for which it can be adapted. For example, the rotary cutter could be a small template used to cut various shapes. Facing operations are possible by adapting the gear rack

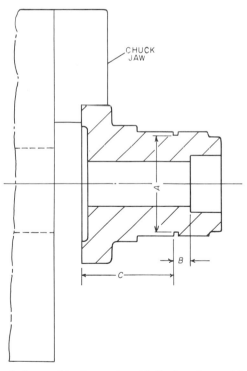

Fig. 42. Housing to be machined on a turret lathe has three critical dimensions: (A) the depth of ring groove, (B) relationship between bottom of the small bore to the back face of the ring groove, and (C) the distance from the back of the groove to the rear face of the housing.

to the cross-slide of the lathe. Also, for straight finish turning, a plain rotary cutter given a counter-clockwise rotation (reversing the position of the rack) will yield improved cutting action and eliminate the thread-like appearance found on many turned parts. As this type of cutter presents more usable surface, even wear and extended tool life are by-products.

Tool Locator Plug Simplifies Turret Lathe Setup

A housing, Fig. 42, machined on a turret lathe in a three-jaw chuck has three critical dimensions: *A*, the diameter of the ring groove; *B*, the depth of a bored hole (in relation to the groove); and *C*, the distance from the face of the flange to the groove. Setting up this job to the required close limits for a new run always presented a time-consuming, vexing problem. The operator spent considerable time and effort with

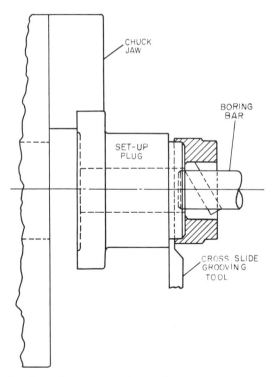

Fig. 43.   Setup plug shows how setting of cross-slide grooving tool and boring-bar are simplified. The broken lines on the right of the sketch are the phantom view of the housing in Fig. 42 and are included to show how the critical relationships of the part are satisfied in this cost-cutting tool.

trial-and-error methods when setting the cutting tools to achieve the correct dimensions in the finished housing.

Figure 43 illustrates the simple, yet effective, solution to the problem that cut setup time to less than a fifth without impairing necessary part accuracy. A round machine-steel plug was machined as shown. Its outside diameter was made the same as the desired outside diameter of the part. The front face was ground to the correct distance from a step (also face-ground) used for setting the grooving tool. In use, the operator merely chucked the plug, advanced and set the boring-bar with the front face of the tool bit touching the ground face of the plug. The grooving tool was positioned and set on the stepped shoulder of the plug.

CHAPTER 6

Special Operations on Lathes

Lathe Tracer Attachment Turns Quill Taper and Grooves

The textile industry purchases aluminum quills, Fig. 1, for various yarn-winding purposes. To simplify the contour-turning of this tapered and grooved part, a tracer and box tool attachment, Fig. 2, was substituted for the regular cross-slide on the carriage. With this device, the quills are contour-turned in one pass from a template. By changing the templates, the number of grooves, or the length, may be varied.

The lathe is a standard economy-model machine, but it has a 1-hp motor to turn the spindle at about 5000 rpm. To drive the parts, a special knurled center was used in the headstock of the lathe and a ball-bearing live center in the tailstock. By pressing the tube firmly over the knurled portion of the headstock center Q, no trouble was encountered in driving the work.

The fixture, Fig. 2, consists of a casting A fitted to the regular carriage cross-slide dovetail ways. Block B is bolted in place, and it carries adjustable steadyrest shoes O and D. These have carbide faces that contact the quill just ahead of the cutting tools. A slot E carries a cutting tool F and is a sliding fit in block A. This slide is held in place by a cover plate G through which, and at right angles to the cutting tool-slide, passes a cold-drawn steel bar H. The bar has a portion milled away, and in this recess the template J is fastened. Two blocks (not shown) are clamped to the lathe bed and to these bar H is fastened.

Fig. 1. Aluminum quill is tapered and has nine grooves, all generated in one pass of the tracer tool.

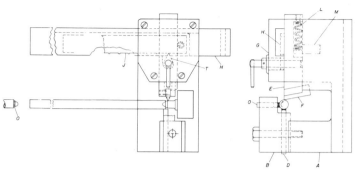

Fig. 2.   Mechanical tracer generates taper and cuts grooves in textile machine quill, Fig. 1.

The template hinges on a tight-fitting screw at one end and has a slot for a clamping screw at the other end. This slot permits swinging the template for fastening at the desired taper. The template is made of flat, hardened and ground stock. A swinging follower has a stylus T which rides the form surface of the template. The follower is held against the coil compression spring L in a slot in slide E, pressing against one end of the slot and loaded against flattened pin M carried by block A.

The illustration shows the cutting tool ready to start a cut. As the carriage moves, the follower rides along the edge of the template, which is a negative of the form that will be cut in the aluminum quill. At the finish of the pass, the feed of the lathe automatically stops. The carriage is returned manually, with the follower swinging away from the template. It is swung into place again at the start of a fresh cut. Time of the operation is about one-half minute. Shoes O and D are carbide-tipped steady-rests to take the cutting thrust of the tool. They are fastened to an extension of the base.

Fixture for Cutting Worm-Gears in a Lathe

Worm-gears can be cut in a lathe with a single-point tool mounted in a boring-bar and held between centers. When using this method, the blank is secured to the compound rest by the device illustrated in Fig. 3. While not recommended as a manufacturing process, this fixture may prove useful in an emergency if no hob or mill is available.

Baseplate A, keyed on the bottom to fit the compound rest B, is secured in place by means of the screws C and clamping bar D. A boss E is tack-welded in position and tapped to receive index-screw F. The baseplate is then bored to a close turning fit with the lower end of

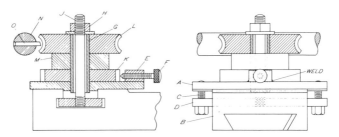

Fig. 3.   Fixture secures worm-gear blank *L* to the lathe compound. Teeth are cut by single-point tool *O* mounted in boring-bar *N*.

bushing *G*. A nut *H* on stud *J*, tightened after each indexing, prevents the assembly from turning while the gear teeth are being cut.

Index-gear *K* and worm-gear blank *L* are made a tight fit on the bushing. However, spacing collar *M* should be a free fit. The index-gear must have the same number of teeth as required on the gear blank or a multiple thereof. The spacing collar is sized so as to position the center of the gear blank at the height of the lathe centers.

In operation, boring-bar *N* is mounted between centers in a lathe with form cutter *O* extended a distance that will give the proper radius to the teeth of the worm-gear. A stop should be placed on the cross-slide, and the depth of cut adjusted by means of the compound feed. As the angle of a worm-gear tooth is 29 degrees, the compound rest can be set at 14 1/2 degrees and the roughing cuts taken on one side and bottom of the tooth space. This will save considerable time.

With the lead-screw of the lathe geared to the pitch required by the worm, cuts are taken by engaging the half-nut and allowing the gear blank to move past the revolving cutter. Several cuts may be taken before indexing for the next tooth. Finishing cuts should be taken by indexing around a second time, after all the teeth have been roughed out.

If the gear to be cut has a lead that can be accommodated by the thread-chasing dial, it may be used. In cutting any other lead the lathe must be left in gear and reversed.

Spiral Chasing Attachment

An arrangement that permits the lathe chasing of spiral grooves is illustrated in Figs. 4, 5 and 6. The radial position of the tool is controlled by a linearly moving cam driven indirectly by the lathe lead-screw. With the proper cams, spirals having either a constant or variable pitch can be produced.

A spur gear *A* (Figs. 4 and 5) is pinned to the lead-screw *B* at the tail-stock end of the lathe. This gear rotates in mesh with a second spur gear *C* keyed to a special drive screw *D*. Gears *A* and *C* are of the same size but drive screw *D* has a lead twice that of the lead-screw. A bearing *E* is clamped to the lathe bed *F* and supports the right-hand end of the drive screw.

Member *G*, essentially a half-nut, is free to pivot on a special screw secured in a supporting plate *H*. This half-nut is engaged with the drive screw by the action of a lever *J*. A spring *K* holds the half-nut in contact with the lever and two cap-screws secure the plate to the lathe tailstock *L*. Bearing *M*, also attached to the tailstock with cap-screws, supports the drive screw in the area below and adjacent to the half-nut. This bearing is bored slightly larger than the major diameter of screw *D* and is slotted at 90 degrees to the bore to receive member *G*. A stop *N* is positioned

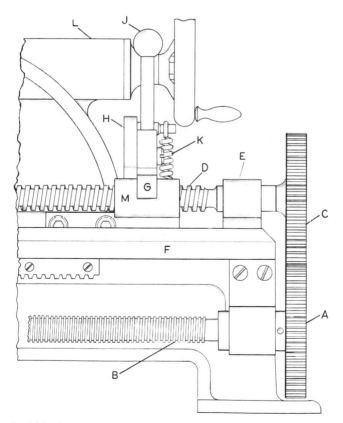

Fig. 4.   Lathe lead-screw *B* is geared to drive screw *D* of the spiral chasing attachment.

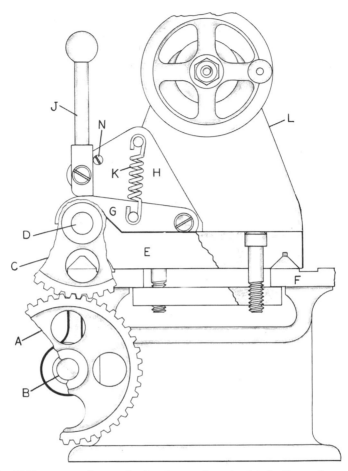

Fig. 5. Half-nut engaging mechanism is seen here in detail. Lever *J* operates the half-nut *G*.

on plate *H* so as to limit the movement of lever *J* when both engaging and disengaging the half-nut. Spring *K* quickly releases the half-nut when the lever is moved to the open position.

Cam holder *O* (Fig. 6) is retained in supporting member *P* by two knurled adjusting nuts *Q*. Member *P* in turn is secured by screws to the rear of the tailstock. Holes are provided in the adjusting nuts for the use of spanner wrenches when setting a cam *R* in its proper longitudinal position. For constant pitch leads, the cam is made in the form of a right-angle triangle having the base equal to twice its altitude, as seen in Fig. 3.

Cam R is attached to the cam holder by cap-screws and dowels, is supported by the lower cross-slide of the lathe, and is held in contact with guide S by a roller follower T. Guide S, made in two parts fastened by cap-screws and dowels, is secured to the rear of the lathe carriage. Roller follower T is free to rotate on a threaded pin in a member U that is attached to the upper cross-slide. The cross-slide feed-screw nut is disconnected and a weighted cable V is used to keep both the follower in contact with the cam and the cam in contact with the guide. Cable V is retained in post W by a screw and passes over a pulley (not shown) attached to guide S.

This device was used on a lathe having an 8-pitch lead-screw, and the gear change box set up for cutting four threads per inch. Since gears A and C are of the same size, lead-screw A and, consequently, drive screw D make two revolutions for each revolution of the workpiece X. Screw D, having a lead twice that of the lead-screw or 0.250 inch (four threads per inch), will, therefore, advance the tailstock and the cam 0.500 inch as the workpiece rotates once. Since the cross-slide is moved toward the operator at twice the rate the cam is advanced by the drive screw, tool Y will cut a spiral groove with an 0.250-inch lead.

Thus, spiral grooves having many other pitches can be obtained simply by changing the gears on the lathe to those that would ordinarily

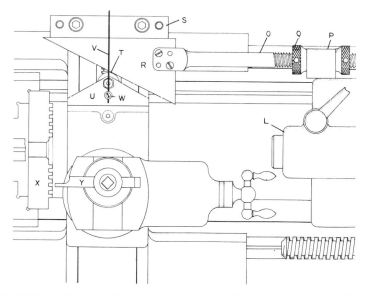

Fig. 6.   Tailstock L, cam holder O, and cam R are advanced toward the headstock by engaging the half-nut with drive screw D. Cam R causes the tool Y to chase a spiral groove by moving the cross-slide against the restraining force of weighted cable V.

be used to cut a thread of that pitch required. If the lathe on which this device is to be used does not have an 8-pitch lead-screw, the same results are obtained by giving drive screw D a pitch double that of the actual lead-screw. Although one cam is used for constant-pitch spiral grooves, accelerating and decelerating leads may be cut using cams with a curved face. Layout of such cams is a simple operation and can be based on the one cam employed for cutting constant-pitch spirals. When chasing a

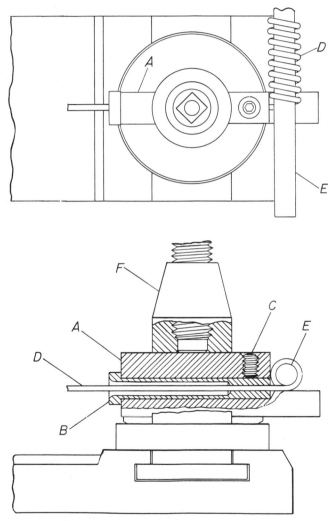

Fig. 7.   Convenient toolpost attachment that facilitates the lathe winding of coil springs.

spiral groove having an accelerated or decelerated lead, a stylus with a rounded but relatively sharp nose is used to replace the roller follower.

The carriage is locked in place during chasing and the compound slide is positioned parallel to the ways. Depth of cut is set by means of the compound feed dial. Also, the threading dial is employed in conjunction with lever J in the same manner as the half-nut lever would be used during a normal threading operation.

Handy Arrangement for Lathe Winding of Coil Springs

A toolpost attachment for the lathe winding of coil springs from various sizes of wire is illustrated in Fig. 7. With this device, the required size and pitch of coil springs can be obtained with a relatively high degree of accuracy.

Holder A is made of steel and is drilled to receive a split bushing B. Pressure-screw C clamps the split bushing on the spring wire D as it passes through the bushing and coils around the mandrel E. Depending on the application, split bushing B can be made of casehardened machine steel or brass. One bushing can be used for several different sizes of wire. The lower side of the hole in any size bushing should line up with the horizontal face of the holder in contact with the wire.

In operation, the toolpost F is adjusted so that the wire is held in continuous contact with the holder at the two points shown. In this manner, equal tension is maintained on the spring wire during the winding operation. Pressure-screw C is tightened to apply the proper feed tension.

Tensioning Holder for Coiling Spring Wire

The hand tool in Fig. 8 is used in making tension or compression springs for winding the wire on a mandrel rotated in a lathe. It insures uniformity of coil size and spacing, and eliminates the cuts and bruises often inflicted on the hands of operators performing this task. In addition, it provides a means of applying the constant degree of pressure necessary to insure the proper tension on the wire, and simplifies controlling and guiding the long free end of the wire.

The tubular steel body A, which is about 1 inch in diameter, has a flat B machined on one side at the right-hand end, to which is brazed or welded a slotted mild steel shackle C. Fitting freely into the slot in this shackle is a steel tension arm D, the right-hand end of which is reduced in width to suit the slot. A fulcrum stud E anchors the tension arm and

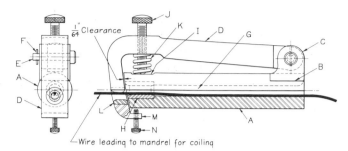

Fig. 8.   Simple hand tool that facilitates coiling wire on a rotating mandrel in producing tension or compression springs.

permits it to swivel. The stud is prevented from moving in an endwise direction by the stop-pin *F*.

The bore *G* of tube *A* may be any convenient diameter, but to insure adequate capacity for the tool it was found advisable to make this diameter not less than one-quarter the outside diameter of the tube member. The mouth of the bore at the right-hand end of the tube is chamfered, and at the opposite end it is enlarged concentrically to provide a tight press fit for the hardened steel bushing *H*. The bore of this bushing is smoothly polished, and is slightly smaller than the diameter of hole *G*. Each end is well rounded in order to prevent any interference from occurring when the wire is being passed through the tube *A*.

A shallow flat *I* on the outside of the tube is in axial alignment with the shackle member *C* on the opposite end. A knurled screw *J*, threaded through arm *D*, abuts against this flat and compresses a coil spring *K*, which exerts pressure against arm *D*. The curve on the inside face of the bent portion of arm *D*, measured from the center of the fulcrum, provides a clearance of approximately 1/64 inch between that face and the formed end of the tube *A*, which has a similar curve. The end of the hardened bushing *H* is also formed to exactly the same radius.

A hardened and polished steel guide bushing *L* is pressed into the lower end of the bent portion of arm *D*. The guide bushing is located on the same lateral center as bore *G*, but is aligned vertically with the bore only when arm *D* is in its highest position relative to tube *A*.

At the extreme lower end of arm *D* is a step *M* through which a small check-screw *N* passes. The step is far enough below tube *A* to permit arm *D* to swing radially about fulcrum *E* for a distance slightly greater than the diameter of the hole in bushing *H*.

In using this tool, the spring wire is freely threaded through bore *G*, from right to left. The adjusting screw *J* is then released to allow arm *D* to move sufficiently so that the wire can be passed through bushing *L*.

After the foremost end of the wire has been gripped in the lathe mandrel, ready for coiling, the arm D is pushed down, thus providing a tension on the wire, which is regulated by the setting of screw J. In this position, the bores of bushings H and L will be out of line a sufficient degree to provide a certain amount of friction on the wire.

The tool is held in this manner throughout all the subsequent coiling of the wire on the rotating mandrel. Once this operation has started, the operator will quickly discern whether sufficient tension is being maintained, and adjustments can quickly be made to increase or diminish the tension by moving screw J a slight amount.

Should it be necessary to stop the coiling operation for any reason before the spring has been completely wound, the device can be gripped tightly to the wire merely by turning screw N until it contacts the side of tube A. The hand gripping pressure on arm D may then be relaxed without fear that the arm will move away from the tube and release tension on the wire. Having thus gripped the wire in the holder, the whole device can be slipped behind the toolpost to prevent the wire from uncoiling on the mandrel.

This device will also prove very useful for straightening out kinks and bends in the spring wire, and also for straightening wire used for other purposes. These kinks will automatically be corrected by merely passing the wire through the two offset bushings H and L. Since passing the wire through such a device in no way damages the polished surfaces, the tool can often be used in cases where straightening cannot be performed quickly or safely by conventional means. If bore G is of sufficient size, the holder can be employed successfully for a large variety of wire diameters.

CHAPTER 7

Special Tooling for Screw Machines

Taper-generating Tools for Screw Machine

Cylindrical parts like that in Fig. 1 were required in production quantities. The available screw machines and turret lathes had no taper-generating tools. Therefore, it was necessary to develop a satisfactory cross-slide attachment (Fig. 2) that would be effective, inexpensive, and offer an easy means of setting the tools; that would keep production and tool costs low; and which would also yield the specified accuracy.

Boring-bar *A* with the cutter *B*, mounted as in section T-T, slides in a channel slide holder *C*. Feed on the bar is delivered by flat-faced plug *D*, pushed by the turret against follower roller *E*. Cover plate *F* retains boring-bar *A*. It is spring-loaded by compression spring *G*, which lies in a slot against pin *H* to return the bar.

The slide mounts on a welded pedestal, clamped to the screw machine cross-slide by T-nuts, as at *J*. The pedestal is positioned by keys *L*, which fit the T-slot. Resting on the pedestal, the tool-slide pivots on pin *M*. The top plate of the pedestal has segment slots for studs *N*, which lock the slide in place at the correct angle for turning the taper.

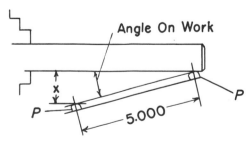

Fig. 1.   Typical taper-generating problem solved by the lathe tooling shown in Fig. 2.

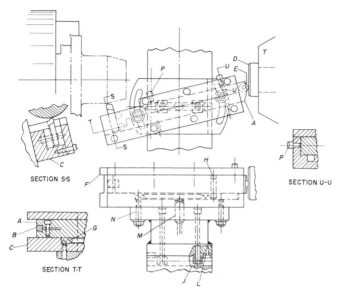

Fig. 2.   Attachment for lathe cross-slide provides adjustment for generating a convenient range of commonly used tapers.

When setting the attachment to turn a required taper, a test bar is first trued on the machine chuck. At two places along the slide channel are reference studs P. The rear stud is brought into contact with the test bar, as in Fig. 1. Then the angle of the slide channel is adjusted to suit the predetermined dimension X. The convex surfaces are spaced a distance apart that is convenient for rapid computation — in this case, 5.000 inches. To determine dimension X, it is only necessary to find the sine of the part angle on a logarithm table, move the decimal point one place to the right, and divide by 2. The dimension can also be found in a table of sine-bar constants, (for example, page 1368, MACHINERY'S HANDBOOK, Seventeenth Edition). A convex wheel was used for grinding the radius on studs P at the 5.000-inch spacing after assembly in their mounting holes, section U-U. Consequently, the original drill spacing need only be approximate.

Profiling Tool for Automatic or Hand Screw Machines

The profiling tool illustrated in Fig. 3 was designed to produce a "clevis terminal," shown by dot-and-dash lines, on a No. 2 Brown & Sharpe automatic screw machine. When the tool has been indexed into position, the front-slide cam of the machine advances the guide piece A

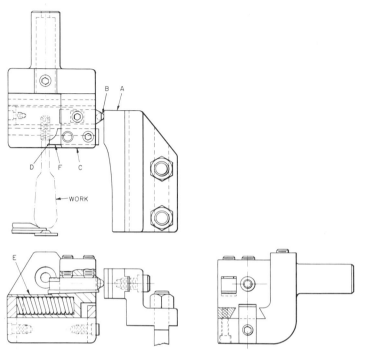

Fig. 3.   Profiling tool for producing formed parts in hand or automatic screw machines.

into contact with the frictionless roll *B*, which operates the tool-carrier slide *C*. The carrier-slide is advanced to a predetermined depth, indicated at *D*, and the tool is fed forward by the lead cam. The pressure of spring *E* causes roll *B*, slide *C*, and the cutting blade *F* to follow the required contour.

In this particular case, the lead cam dwells, so that the head of the part is formed by advancing the tool with the front cross-slide. A telescopic support bushing may be used in the tool shank.

End-stamping Tool for a Screw Machine

Brass screw-machine parts required a code stamp on one end. In order to eliminate a separate stamping operation, the tool illustrated in Fig. 4 was designed to stamp the part while rotating in the machine and before cutting off. An expanding driver is inserted in a drilled hole in the part (as it rotates) to drive the stamp at spindle speed. This saves the time that would otherwise be necessary to stop and then start the spindle again when moving the stamp forward.

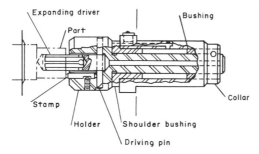

Fig. 4.   Device for end stamping of screw-machine parts without stopping the spindle.

The expanding driver has four slots through the spring-tempered end, and the four resulting prongs are spread enough so that they are friction-tight in parts with a maximum-size hole. This enables the stamp to rotate at spindle speed. The driver has about 0.010-inch float in the holder and is driven by a cross-pin. This pin is pressed in the holder and goes through a clearance hole in the driver.

End thrust is taken by the shoulder of the front bushing, and both bushings are equipped with oil-grooves. In addition, the sleeve that fits in the turret is provided with an oiler and suitable oil-holes. The end of the holder is left soft in order to permit pinning of the retaining collar in place as illustrated in Fig. 4.

Simplifying the Tooling of Hand Screw Machines

The hand screw machine is convenient for making parts in quantities insufficient to warrant the use of an automatic screw machine. A variety of tooling is available that enables all stations of hand screw machines to be employed for both single and combined operations. When properly set up, these machines are comparatively simple to operate, the key to best use of the equipment being tooling which makes setup fast and accurate.

A good way to assure rapid and precise setups for repetitive jobs is to retain a machined bar end piece when finishing a lot of parts. This piece can be placed in the collet and used to set the tooling when doing the job again. For initial setups of cross-slide plunge-cut tooling, the tool-setting fixture shown in Fig. 5 has been found useful.

The fixture consists of an accurately ground bar A to which a dial indicator B is attached. The bar, in turn, is held in a collet and adjusted so that flat C is perpendicular to the cross-slide D. This is accomplished

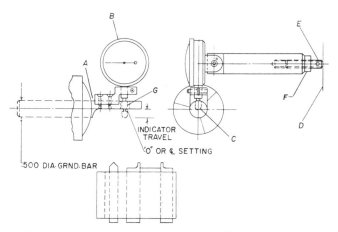

Fig. 5.   Fixture that speeds accurate setting of tooling on hand screw machines.

by means of adjusting screw *E*. Once the flat has been adjusted to the vertical position, the adjusting screw can be locked by nut *F*. The adjustment will then remain fixed for future setups in the same machine.

The dial indicator must be zeroed to the center line of the spindle. This is done by placing a flat-surfaced piece or gage-block against flat *C* and then moving the indicator point *G* against the gage-block to the center-line position. Next, the indicator is advanced until the needle has moved approximately three-quarters of its total travel. At this point the dial is set at zero and the indicator is locked in place.

Tools can be set in relation to the center line of the workpiece by advancing them against the indicator point until one-half of the diameter to be cut is read on the dial. The cross-slide stop is then adjusted for this tool position. All succeeding cross-slide tools can be positioned according to the established indicator (center-line) setting in the same way.

The adjustable tool-holder seen in Fig. 6 will serve well with the setting fixture here described. It is a modification of commercially available tooling. Illustrated is a combination of accurately sized tool bits and spacers arranged to provide a fast, economical setup for machining groves in shafts for rubber O-rings. In practice, this tooling has been found adequate for many shaft-grooving operations.

The multiple tool-holder *H* is designed for attachment to the cross-slide of the machine and is large enough to accommodate 3/8-inch tool bits. It is important that the tool bits have very accurate width dimensions so that accumulation of size inaccuracies is minimized. The necessary tool bits are inserted in the holder, and spacers *J*, as required, are inserted between them.

Clamping screws K are first tightened sufficiently to apply only a slight clamping force against the side of the stacked tool bits and spacers. Clamping screws L are then also tightened to apply a slight clamping force on the top of the stacked tool bits — but not on the spacers. Next, tool-setting screws M are adjusted to position the tool bits so as to produce the required part diameter. At this stage of the setup, the tool bits can be positioned relative to the center line of the part by use of either the tool-setting fixture (Fig. 5) or a bar end piece. After all tool bits are correctly positioned, the clamping screws are fully tightened, except for those screws L located over the spacers. The multiple-tool setup is then complete.

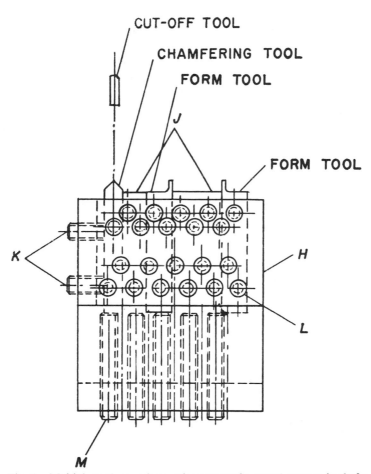

Fig. 6.   Multiple-tool setup for cutting accurately spaced grooves in shafts.

The tool shown in Fig. 7, when provided with spacers N and cutting tools as accurately sized as those used for the external grooving setup, can provide a quick versatile means for producing internal grooves. The internal grooving tool is designed for attachment to the cross-slide; adapter P is made to fit the equipment to be used. In addition, the tool can be used in adjustable boring heads for internal grooving on a milling machine or a jig borer. The cutters and spacers are ground to accurate thickness tolerances as before and are stacked to obtain the required dimension X between the grooves. Cutters Q are similar to standard circular cutoff tools, and all spacers and cutters are keyed to the arbor R to prevent rotation during the machining operation.

Since the key establishes the relationship between the cutting edge of the tool and the arbor, the cutters must be sharpened the same amount relative to the key slot if the tolerances on the depths of the grooves are to be closely held. If this is not done, the cutting edges will not be in the same plane relative to the center line of the part and the depth of the grooves will not be equal. The cutter relief will also affect the groove depth if an equal amount of material is not ground from the face of the cutter.

The best way to prevent unequal grinding of the cutter faces is to sharpen the cutters while they are on the arbor. This can be accomplished by clamping the arbor either in a V-block or between centers with the cutting surfaces parallel to the grinding machine table so that the faces can be ground simultaneously.

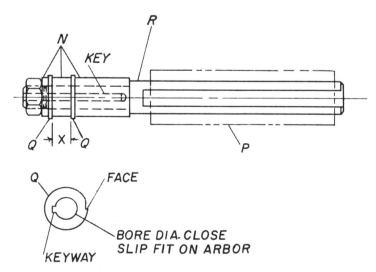

Fig. 7.   Tooling arrangements, similar to that in Fig. 6, for producing internal grooves.

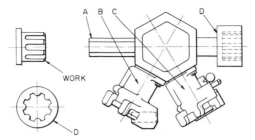

Fig. 8.   Plan view of an unusual turret lathe setup for machining a plastic part.

Plastic Knobs Produced in Small Lots on Hand Screw Machine

Plastic knobs for radar equipment were successfully produced on a hand screw machine when the limited quantity required made the cost of molding prohibitive. The complicating factor was that a series of shallow flutes to give finger hold had to be cut along the outside diameter of the knobs. Figure 8 is a plan view of the set-up of the hexagon turret of the machine.

The over-all length of the knobs, made from 7/16-inch Tenite No. 2 stock, was controlled by a stop-bar *A* located at the first of the four hexagon turret stations used for the job. A cutter *B* at the second station was used to produce a chamfer at the end of the knob, then the smaller diameter was turned by means of a box tool *C* supported in the third station.

To form the grooves, a simple cutting die *D* of cold-rolled steel was designed. The die was supported in the fourth station, and for this operation the motion of the machine spindle was, of course, arrested while the turret ram was advanced and retracted. After the knob was cut off, a shaft hole was drilled in the flange end in a secondary operation.

CHAPTER 8

Special Tooling for Drilling and Associated Operations

Tool Chamfers Both Ends of Holes Simultaneously

Drilled holes in boiler tube sheets required chamfering on both ends. The holes were machined on a radial drill equipped with a quick-change chuck for holding the drill, reamer and chamfering tool. The sheets were supported on parallels. After the ends of the holes on one side of the sheet were chamfered, the sheet was turned over to chamfer the ends on the other side. Since the sheets were awkward to handle and heavy, production costs were high. The solution to this problem was the development of the tool illustrated in Fig. 1. This tool chamfers both ends of the holes simultaneously.

The cutter, which is mounted in a pilot member, is pivoted. Its position is controlled by an adjustable rod that has a 45-degree chamfer and a set-screw flat on one end. A cone-pointed set-screw adjusts the rod. The cutter is retracted against its stop by a spring pin.

A collar carrying three ball-housing bushings is pressed on a diameter slightly larger than the pilot. Vertical positioning of the balls is accomplished by a backing disk. A lip on the bushing, as shown in the enlarged view, is rolled over to retain the ball, leaving it free to turn.

Stop for the tool body is a threaded adjusting nut. This is locked by a soft plug and set-screw. The pilot member of the tool is driven by a drive key. Its upper end acts as a stop when spring pressure brings the two members together as the spindle is raised.

On the downstroke of the spindle, the balls contact the part. Further spindle travel brings the cutter into working position. Downward movement of the tool body is stopped when it comes into contact with the adjusting nuts. After a few revolutions of dwell, the chamfers are completed.

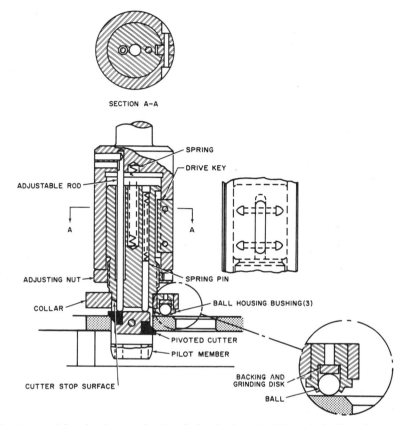

SECTION A-A

SPRING

DRIVE KEY

ADJUSTABLE ROD

A

A

ADJUSTING NUT

SPRING PIN

COLLAR

BALL HOUSING BUSHING(3)

PIVOTED CUTTER

PILOT MEMBER

CUTTER STOP SURFACE

BACKING AND
GRINDING DISK

BALL

Fig. 1. Tool for simultaneously chamfering both ends of holes in boiler tube sheets.

Trepanning Tool of Unusual Design

When cutting 4 1/2-inch diameter holes in the web of steel car wheels, ordinary trepanning tools often broke and required time-consuming grinding and resetting. An unusual trepanning tool was therefore developed which utilizes two cutters and is capable of taking heavy cuts.

There are essentially four main parts to the tool, body A, tightening sleeve B, and cutters C and D, as seen in Fig. 2. The body has a No. 6 Morse Taper and is bored out at the bottom end for lightness. Coarse left-hand threads on the body engage the tightening sleeve which is bored and ground on its lower end to fit the tapered ends of the cutters.

The cutters are seated by fitting shoulders E into recesses in body A and are secured to the body as a tapered conical surface on sleeve B is moved downward on corresponding surfaces on the cutters. This is

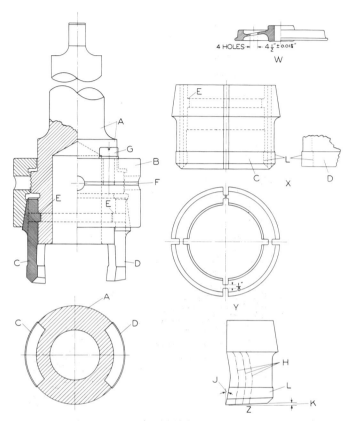

Fig. 2. The heavy duty cutters held by this unusual trepanning tool have different forms in order to reduce the load on each cutting edge.

accomplished by screwing sleeve *B* on the threads of the body. In trepanning, the upward thrust of the cutters is taken by shoulders *E*. To keep the sleeve tightly screwed, it is split at *F* and locked by screw *G*. The cutting edges of the two tools differ in shape in order to reduce the cutting load.

As shown in view Z, a cutter may be reground several times along lines *H*, which extend to as much as one-half the width of the cutter and still have sufficient cutting surface for further service. The front rake angle *J* and the clearance angle *K* are about 15 degrees and 5 degrees, respectively. Section *L* on the cutter is relieved about 0.008 inch on both the inside and outside for clearance. This does not exceed the tolerance of plus or minus 0.015 inch allowed on a hole in the wheel web, as shown in view W.

To simplify making of cutters, two pieces of flat steel stock are heated and rolled into circular shapes with their ends meeting as closely as possible. After being annealed, one ring is machined to the shape seen at C and the other to shape D, as shown in view X. The rings are then slotted into four sections, each section being connected by a 1/8-inch bridge, as seen at Y. Each segmented ring is heat-treated, hardened and tempered, and finally separated into four pieces, thus providing four pairs of cutters.

Special Cutter for Beveling and Chamfering Large Tubes

A countersinking tool for cutting an internal bevel in large tubes to enable them to be held on centers, and at the same time cutting a chamfer on the outside edge of the tubes, is shown in Fig. 3.

The cutter adapter A is made from S A E 4140 bar stock, turned down at one end to receive a special fifteen-fluted cutter B. A pin C, which passes through the cutter and the front shank of the adapter, transmits a positive drive to the cutter. A fine thread D is cut on the outside diameter of the adapter for adjustment purposes when it is desired to vary the bevel dimension. The outer end of the adapter is turned to a Morse taper to fit the machine spindle.

A hardened and ground sleeve F has three functions — as a stock stop, a bearing for a guide bushing G, and a holder for cutter H when it is desired to chamfer the outside diameter of the work J. Two set-screws K position the cutter. Locking collar L holds the adapter in the sleeve. Fixture M contains the guide bushing that accurately centers the tool and rigidly supports the cutter. Rapid removal of the cutter is accomplished by loosening the socket-head cap-screw N.

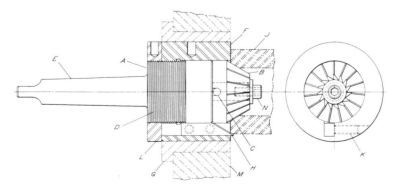

Fig. 3.   A combination countersinking cutter and chamfering tool for tubing.

Method for Off-Center Drilling of Cylindrical Parts

Off-center drilling through cylindrical parts in the manner illustrated in Fig. 4 is difficult enough under the most favorable conditions — such as when drilling close to the center line. The farther the hole from the center line, the greater the tendency of the drill to "walk" or be deflected down the incline so that it deviates from the intended path. The result will be a curved hole, as indicated by the broken line at A; the smaller the drill, the more pronounced the curvature. Fine drills may break, and even if they do not break, there will be inordinate wear both on the drill and in the bushing, aggravating the initial tendency to "walk."

The drill can, however, be started true by the method shown in Fig. 5, which consists of first spotting a slight flat on the workpiece with a fishtail or toothed end cutter. The diameter of the cutter should be just large enough to overlap the hole to be drilled, with its radius enough larger than the hole (as shown by the inset at B) to permit starting the

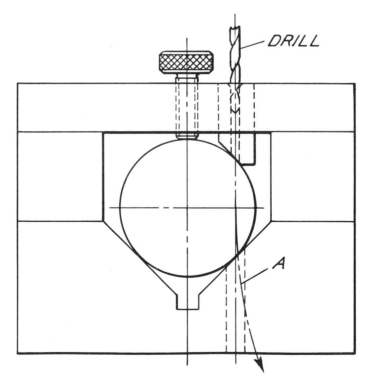

Fig. 4.   Simple jig for use in drilling off-center hole in cylindrical part. This jig has the disadvantage of allowing drill to be deflected from straight path.

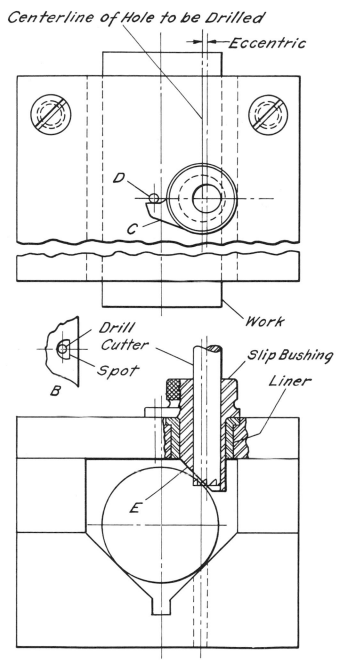

Fig. 5. Drill jig designed to overcome difficulty experienced in drilling a straight off-center hole with jig illustrated in Fig. 4.

drill on a flat surface. If, for example, the drilled hole is 1/8 inch in diameter, then the cutter should be about 5/16 inch in diameter.

Two slip bushings are required which are interchangeable in a liner. One bushing is made concentric, for the drill. The other bushing is eccentric, the amount of eccentricity being, say, 5/32 inch for the drill-to-cutter ratio previously stated, or proportional to the drill and cutter used. Both bushings have a tail C which comes in contact with a stop-pin D, to prevent rotation of the bushing. The lower ends of the bushings are beveled to suit the contour of the workpiece for tool support, illustrated at E. As the stop acts in the direction of spindle rotation, the bushings will "stay put" during spotting and drilling. Depth of cut can be controlled by a stop-collar or by the regular drill stop.

Spotting the cylindrical piece as described provides a flat surface on which the drill can be started. Then by using a light feeding pressure it is possible to produce a straight hole in the work with even the smallest drill in the regular wire gage sizes.

Internal Grooving Tool with Positive Depth Control

An inexpensive, yet efficient, internal grooving tool developed for use on lathes or drill presses is shown in Fig. 6. This tool has a simple adjustable means for accurately producing annular grooves of various depths in the bores of workpieces.

The device is assembled on a spindle A, the shank end B of which is ground to the taper of the machine quill in which the tool is to be used. Mounted on spindle A is a flanged housing C, having an outside diameter slightly less than the blind bore in workpiece X, in which annular groove Y is to be machined.

Integral with flange C is a hub D that is bored concentrically to be a close sliding fit over the parallel portion of spindle A. Flange C is pre-

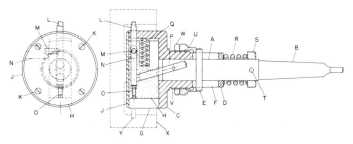

Fig. 6.   Internal grooving tool with an adjustable stop for controlling the depth of grooves machined in the bores of workpieces.

vented from rotating on the spindle by driving pin E that is press-fitted into a hole drilled diametrically through the spindle. Each projecting end of the pin passes into elongated slots F which are machined through opposite sides of hub D.

A deep rectangular slot G is machined centrally in the left-hand end of flange C so as to pass through one side of the flange and terminate close to the edge of the other side. A close-fitted rectangular steel slide H is set to move smoothly within this slot. The slide is held against the bottom of the slot by a thin steel disc J which is affixed to flange C by four countersunk flat-head screws K.

The cutting tool L has a long cylindrical shank that fits closely into a drilled hole extending the full length of slide H. A hole in the wall of flange C provides additional support for the front of the tool. The cutting tool can be fixed at any required radial setting by headless screw M, located in the side of slide H.

An elongated slot N is machined through the adjoining wall of flange C to afford access for adjusting and locking screw M. Slot N is sufficiently long to clear the projecting end of this screw at all points throughout the normal traversing movement of slide H. Cutting tool L can be advanced or retracted by means of set-screw O in the lower end of the slide. The set-screw has fine pitch threads to permit making precision adjustments.

Pin P is press-fitted and doweled in an angular blind hole in spindle A. The opposite end of this pin is a sliding fit in another angular hole in slide H. The lateral movement of slide H is derived from the pressure exerted on this member as pin P and spindle A are advanced and retracted. A blind hole is machined in one end of slide H for compression spring Q, the opposite end of which bears against the wall formed by slot G in the flange. A second coil spring R fits freely over spindle A between hub D and collar S. The collar is permanently fastened to the right-hand end of the spindle by dowel T.

Member U is a steel adjusting-ring that screws over the threaded portion V of hub D. A threaded lock-ring W is used to position adjusting-ring U according to the depth of the groove required. By altering the longitudinal setting of ring U, the lateral movement imparted to slide H, and thus the amount of penetration of cutting tool L into the workpiece, can be controlled as required.

In operation, the grooving tool is fed by the tailstock spindle into the bored hole in the workpiece until disc J bears against the back face of the bore. Continued inward movement of the tailstock spindle causes the spindle A to slide into flange C. Springs Q and R are compressed, forcing pin P farther into the hole in slide H and thereby moving the latter gradually outward. Cutting tool L, affixed in the slide, is therefore advanced and commences cutting into the surface of the bore to form groove *Y*. This cutting action continues until spindle *A* has passed sufficiently into the flange to bring driving pin *E* into contact with the end of ring *U*, as illustrated in Fig. 6.

After the groove is machined, the tailstock spindle is retracted, whereupon springs *Q* and *R* expand, holding flange *C* in the bored hole, and slide *H* and cutter *L* withdraw into the flange housing under the impetus of pin *P*. Spindle *A* continues to move to the right until pin *E* bears

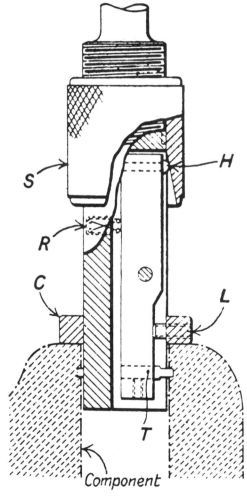

Fig. 7.   Recessing tool *T* is fed into the work when projection *H* is depressed by the conical surface in sleeve *S* as the sleeve moves downward.

against the back ends of elongated slots *F*. The complete device will then take up the retracting movement of the tailstock spindle, and flange *C* will be withdrawn from the bore in the workpiece.

Recessing Tool Designed for Use on a Drilling Machine

The recessing tool illustrated in Fig. 7 was designed for use on a drilling machine. The component to be recessed is a spindle bracket casting for a die-sinking machine, and its size would have necessitated the use of a large lathe had the normal method of mounting been adopted. Since no other turning was necessary, it was decided to carry out the recessing operation on the same drill press as was used for drilling and reaming the two main bores in the casting.

Recessing tool *T* is mounted at the lower end of a pivoted lever, and means are provided for feeding it into the work while the shank is rotating. The feeding movement is produced by lightly gripping the knurled sleeve *S* by hand and retarding its rotation. This causes the sleeve to move downward on a fine-pitch, left-hand thread on the body. The conical surface in the bore of the sleeve skirt gradually depresses projection *H* as the sleeve moves down, thus feeding tool *T* out.

The distance of the groove being recessed from the top surface of the casting is determined by collar *C*, in which there is an adjustable set-screw *L* which limits the outward travel of the tool and controls the depth of the groove. When the sleeve is returned to its upper position, the tool is retracted from the groove by the action of spring *R*.

Very slight pressure is required in gripping the sleeve, and increased resistance indicates that the tool needs sharpening.

Drilling Setup for Closely Spaced Holes

The drilling of closely spaced holes in a part often presents a problem when it is desirable to drill two or more holes at one time. This was the case with the cold-rolled steel shaft shown at the right in Fig. 8. Two small holes must be drilled through the center.

The closeness of the hole center distance made it impossible to drill both holes at one time with standard drilling equipment or even with a special two-spindle drill head. Since it was a high-production part and manufacturing costs had to be kept to a minimum, two separate operations were out of the question.

The problem was simply solved by devising the setup shown with which the two are drilled at the same time from opposite directions. The part is nested in a hardened and ground V-block *A* which was screwed

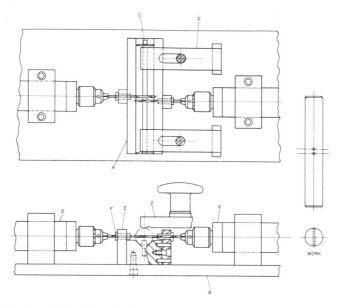

Fig. 8. Tooling designed for simultaneously drilling two closely spaced holes through a shaft.

and doweled to base *B*. Dowel-pin *C* serves as an end-stop and two strap clamps *D* hold the piece securely in the vee. Two plates *E* are bolted and doweled to either side of V-block *A*. These plates carry drill bushings *F*. Two standard air feed drills *G* perform the drilling. While the drills are feeding, the operator is busy removing burrs from a previously drilled part.

Device for Drilling Two Small Holes Close Together

In a production job it was necessary to drill two 5/64-inch holes close together and for economical reasons it was desirable to produce both holes simultaneously. The unique device illustrated in Fig. 9 was designed to meet these requirements. It can be driven from a conventional drill chuck.

The various components are mounted on an aluminum plate. There is a central drive-shaft that is supported in two bronze bearings and carries a tapered neoprene roller at the lower end. This roller is mounted between two washers and nuts in such a way that it can be adjusted axially to provide good contact with two driven rollers on individual spindles. The drive-shaft is retained axially on the aluminum plate by collars and bronze washers.

The driven rollers are also made of neoprene. The shafts on which they are mounted run in bronze bearings and are also provided with ball thrust bearings to take end thrust as the drills are fed into the work. Each driven shaft is drilled and tapped to receive a set-screw which is tightened against a flat on the end of the corresponding drill shank to securely lock the drill to its driving component.

The shanks of drills selected for use with this device are turned down for a short distance and soldered in long extension shanks. The shank extensions should be made of drill rod so as to provide for long life,

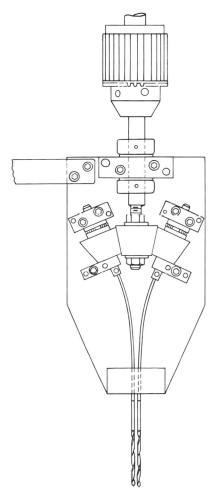

Fig. 9. Unique device designed for simultaneously drilling two small holes close together.

there being a constant reversal of stresses as the drills whip around during rotation. Extra drills soldered on extension shanks are kept on hand for quick replacement. The entire unit can be enclosed with a cover made of sheet metal or plastic sheets. A bar fastened to the aluminum plate projects from the device to the column of the drill press and thus prevents the unit from rotating with the machine spindle during an operation.

Either the rotation of the drilling machine must be reversed from conventional practice in using this device or else left-hand drills must be employed. If the device should be designed for larger drills than 5/64 inch, the height of the unit must be increased somewhat so as to avoid too sharp a bend in the drill shanks.

Low-Cost Counterboring Done with Drill Press

Exact depth of counterboring cuts is assured in the shortest time through the use of a novel arrangement using a drill press for short-run jobs. A flange is turned for the shank of the counterboring tool. After press-fitting to the shank, the relieved underside of the flange is finish-turned.

In contact with this lower surface is the anvil of a dial indicator which is mounted on a bracket stand as shown in Fig. 10. Using this arrange-

Fig. 10. Flange and indicator arrangement is simple and inexpensive for short runs. It obviates the frequent need for stopping the machine and withdrawing the tool to measure depth.

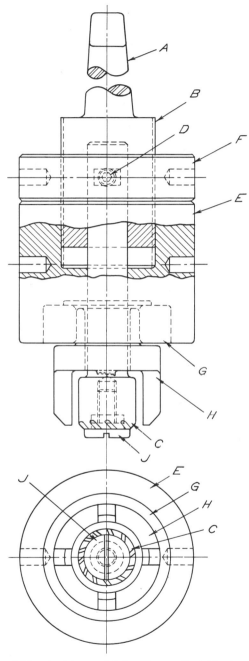

Fig. 11.   Countersinking tool is provided with an adjustable stop for positive depth control.

ment, the operator watches and accurately controls the depth of cut by observing the indicator as the counterbore feeds into the bore.

Countersinking Tool Combined with an Adjustable Stop

When large quantities of parts require countersinking, the problem of accurate depth control often arises. This may be simply met by the tool illustrated in Fig. 11, having a positive self-contained depth stop.

Integral with tapered shank A is a cylindrical body B. Fine-pitch threads are cut around the periphery of the body portion. A blind hole is drilled into the threaded body along its longitudinal axis to receive the straight shank of countersinking tool C. The tool is held in place by two set-screws D that bear against flats machined in its shank.

Threaded on B are stop-body E and lock-nut F, both members being provided with spanner holes. A roller thrust bearing G is pressed into a counterbore in the lower face of the stop-body. The shank of stop H is pressed into this bearing and is free to turn with it. The counter-sinking tool turns freely within both the stop-body and the stop.

The head of screw J which is threaded into a tapped hole in the cutting end of the tool, serves as a pilot. A shoulder on this pilot screw fits within a comparatively deep counterbore, thus providing ample clearance for tightening in the event of tool grinding.

In operation, the desired depth is set by backing off lock-nut F and adjusting the stop-body to obtain the proper distance from the bottom of the cutting edge to the lower surface of stop H. The pilot is then located in the hole to be countersunk, and the tool advanced. When the required depth has been reached, the stop will contact the surface of the work, thus halting the progress of the countersinking tool.

High-Speed Attachment for Large Drilling Machines

An attachment providing high speeds and a sensitive hand feed for large drilling machines not originally equipped with these features is illustrated in Fig. 12. By this device, heavy and unwieldy workpieces that require holes of both large and small diameter can be drilled on a machine of this type, and thereby eliminate the necessity of setting up the part twice.

The spindle of the drilling machine drives the mechanism by means of the tapered shank of shaft A. This shaft also supports frame B as collar C and pulley D are both secured in place. Drill chuck E is mounted on the lower end of auxiliary spindle F. Pulley G is keyed to drive this auxiliary spindle but is free to slide along the keyway H. The two pulleys have a

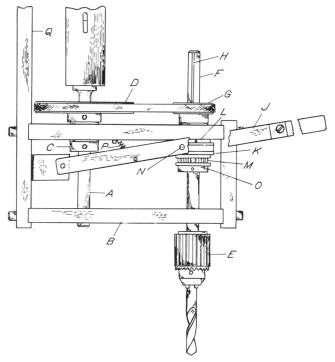

Fig. 12.   This attachment permits the drilling of small-diameter holes located in work usually handled on a large, slow-speed drilling machine.

3 to 1 diameter ratio. A sufficient range of speeds is obtained by varying the main spindle speed.

Forked lever *J* for the sensitive hand feed pivots on the frame at one end and straddles the rectangular collar *K*. Mounted between collar *L* and ball thrust bearing *M*, this rectangular collar is a free fit on the auxiliary spindle and is grooved on two sides to receive the pins *N*. Since collar *L* and collar *O* are both affixed to the auxiliary spindle, any movement of lever *J* is transmitted through pins *N* to collar *K* and the spindle *F*. A spring *P* holds the lever in the raised position.

The frame of the attachment is kept from revolving by the back plate *Q* which extends upward to rest against the arm of the machine. A guard can be added for the pulleys and belt.

Handy Feed-Lever Switch Facilitates Drilling

A micro switch and push-rod attached to the feed mechanism of a drilling machine provide a convenient power control. With such an

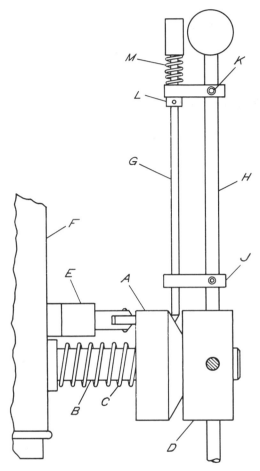

Fig. 13. Arrangement that permits power for drilling machine to be turned on or off without releasing the feed-lever.

arrangement, as illustrated in Fig. 13, a mechanic can start or stop his machine without releasing the feed-lever. The switch is especially useful where much tool changing or work adjustment is required.

Collar A is chamfered on one edge and made a sliding fit on feed-shaft B. Spring C, also mounted on the feed-shaft, serves to hold the chamfered face of the collar in contact with hub D. A micro switch E is secured to the machine column F so that the roller on the plunger is in contact with the unchamfered face of collar A.

A pointed switch-rod G is attached to the feed-lever H by two short blocks J. These are held in place by set-screws K. A pinned collar L

and a spring M keep the pointed rod just out of contact with the chamfer on collar A. Depressing the switch-rod moves both the collar on the feed-shaft and the micro switch plunger to the left. This action is sufficient to turn on the power at the machine. The micro switch is connected parallel to the regular switch.

In operation, the mechanic can feed the tool down, line it up with the work, depress the switch-rod, and start drilling — thus keeping his hand on the feed-lever. There are also many occasions when cutting tools must be changed several times while machining a single workpiece. With this setup, the operator does not have to grope for the switch while substituting tools.

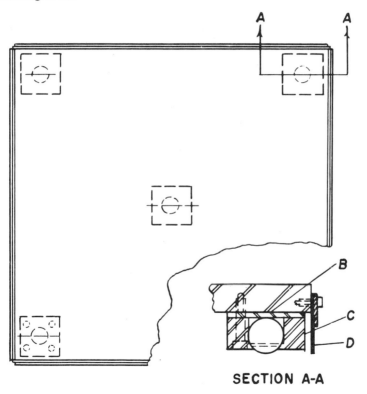

SECTION A-A

Fig. 14.   Ball-bearing movable table designed for use on drill presses.

Easily Manipulated Movable Table for Drill Press

In drilling parts such as heavy castings and slate panels that are awkward to handle on a drilling machine, considerable time can be saved and better work produced if a ball-bearing table is used to move the part in various directions beneath the machine spindle. A table such as the one shown in Fig. 14 will make it possible to align the workpiece to the spindle with very little physical effort.

The table consists primarily of a steel top plate that is ground on both sides. Underneath each corner is a ball-bearing assembly, as shown in Section A-A. There is a similar unit under the center of the plate. Above each ball is a hardened plate B made from flat ground stock. A hole drilled in a mounting block C retains the ball. Strips D, of neoprene, are attached to the four sides of the plate to serve as wipers and prevent any chips from interfering with free rotation of the balls. A wear plate should be mounted on the drill press table to provide an even surface for the balls to roll on.

CHAPTER 9

Jigs for Drilling and Associated Operations

Drill Jig with Bushing Location and Wedge Clamping Features

The jig shown in Fig. 1 was designed to permit accurate drilling and reaming of a hole in an interchangeable part for grinding machines. The part X, a saddle clamping rod post, has a square shank on its left-hand end, and the required hole must be square with this shank. This is accomplished by locating the workpiece in a bushing, and clamping by means of a screw-actuated wedge.

Secured and doweled to body A of the jig is a bushing plate B, into which liner bushing C is pressed. A slip drill bushing D fits into the liner bushing, and is replaced by a slip reamer bushing after drilling the required hole. The slip bushings are locked in place by screw E.

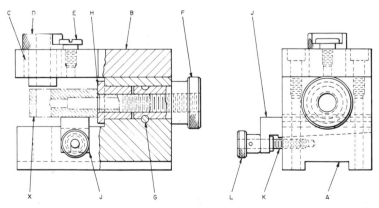

Fig. 1.   Jig employed for drilling and reaming a hole through the square-shanked end of a grinding machine detail X. The workpiece is accurately located and aligned by wedge J and bushing H, and clamped by nut F which is screwed on threaded end of workpiece.

A special nut *F*, placed in a hole in the right-hand end of the jig body, is internally threaded to fit the threaded end of the workpiece. This nut is held in place in the jig (but permitted to rotate) by pin *G*, which fits into grooves machined in the nut and jig body. Pressed into the same hole in the jig body, but from the opposite cut-away end, is a work-locating bushing *H*. The bore of this bushing is a sliding fit for the cylindrical shank on the workpiece.

Wedge *J* is mounted in a beveled slot machined in the jig body. Stud *K*, which operates the wedge by being screwed into the jig body, passes through a slot in the wedge. The stud is screwed into or out of the body by means of knurled knob *L*, which is pinned to the outer end of the stud.

In operation, the workpiece to be machined is slid into the jig from the left-hand, cut-away end, and inserted in bushing *H*. Nut *F* is then screwed on the threaded end of the workpiece that projects through the locating bushing. This action draws the workpiece toward the shoulder on bushing *H*. However, before contact is made with the bushing shoulder, wedge *J* is fed in by rotating knob *L*. The wedge, acting on the under side of the square shank of the workpiece, forces the workpiece into proper alignment. Nut *F* is then tightened to bring the workpiece against the bushing shoulder, and the drilling and reaming operations are performed.

Pin Drilling Jig Uses Toggle Action

For a high-volume production run a hole had to be drilled on the diameter of a bolt's head. A simple welded jig, incorporating a drill bushing, was constructed, enabling the holes to be made with great accuracy at high speed on a drill press (see Fig. 2). View X shows, in partial cutaway, left, the design of the jig with the workpiece stem cavity *W* in broken lines, and the center line of the drill at *D*. *B* indicates the standard drill bushing. Here the jig is ready for loading. The end view at the right shows the clamping bar held vertical, in readiness for loading.

The bushing holder is mounted at the top of a steel block which is drilled with a sliding fit for draw-bar *A*, as well as cavity *W*. The draw-bar is fitted at the front end (left) with a pivoting toggle *T*. The opposite end of the draw-bar carries the clamping head *C*. A compression spring, coiled around the draw-bar, between the block and the clamping head, tends to force the clamping head outward to the right.

When loading the jig the toggle handle *M* is positioned vertically in both planes, as in View X. A workpiece is placed in the stem hole *W*. Then the toggle handle is swung to the right (as shown at *S*), reaching

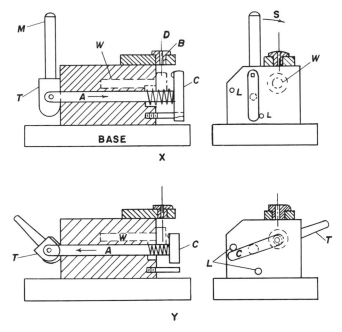

Fig. 2. Toggle drill jig in loading position, view X. Drilling is done with toggle *T* swung to the right, pulling clamping head *C* to the left, into the work, as shown in view Y.

the position seen at the right in View Y. Thus the keeper stud on clamping head *C* will contact the head of the workpiece when the draw-bar is pulled back by means of the toggle handle. This action is illustrated in the partial cross section at the left in View Y. Drilling can now take place.

To remove the work the toggle is released and the handle is swung to the vertical. The rear face of the block has two studs *L*, which limit angular rotation of the clamping head.

Low-Cost Drill Jig Made of Structural Shapes

Not a great deal of accuracy was required in a farm-implement component, and the quantity to be made was not great, but it was still highly impractical to lay out each piece separately for a number of drilling operations. Because of the low production, tooling costs had to be held to a minimum. This presented a problem, since holes had to be drilled in two different planes at right angles to each other.

The problem was solved by the selection of length of standard structural I-beam for the main body of a tumble jig. As may be seen in Fig. 3,

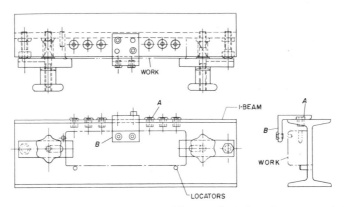

Fig. 3.   Tumble type drill jig constructed inexpensively by using structural shapes.

the I-beam served the multiple purpose of the jig base, upright, bushing plate, and feet for drilling in two directions. Finishing of the I-beam was not necessary to achieve accuracy.

Three standard dowel-pins were pressed into the web of the beam to serve as work locators. Four holes were tapped in the web to receive clamps, studs, and heel screws. The clamps were of the ordinary hand-knob strap type and were purchased complete from suppliers.

Six drill bushings *A* were provided in holes drilled along the top of the beam. Two additional drill bushings were provided in bushing plate *B*, which is simply a piece of standard structural angle iron. The angle iron was screwed to the beam flange by socket-head machine screws and doweled in place. While the design of the jig was most simple, it served its purpose well.

One Drill Bushing Locates Two Holes

Two holes, one off-center and one radial, were drilled through the wall of a cylindrical part by using a jig having a single, rotatable bushing. Separate bushings could not be used because of the close spacing between the holes. The workpiece, approximately 12 inches in diameter, was held on a large dividing head.

Features of the bushing design appear in Fig. 4. From the reference center line X-X of the workpiece, off-center hole *A* is located a distance of 0.3125 inch. Radial hole *B* lies 15 degrees from the center line, in an opposite direction. The rotatable jig bushing *C* has two holes, *D* and *E*, corresponding to the holes *A* and *B* to be produced in the workpiece. Notches *F* and *G* are 90 degrees apart in the bushing periphery.

The bushing position illustrated is for drilling hole A. Next to the bushing is a locator H having a plunger J which engages notch F. To drill hole B, the plunger is retracted by handle K, and both the workpiece and the bushing are indexed in the direction indicated by the arrows — the workpiece, 15 degrees, and the bushing, 90 degrees, so that notch G is then engaged by the plunger.

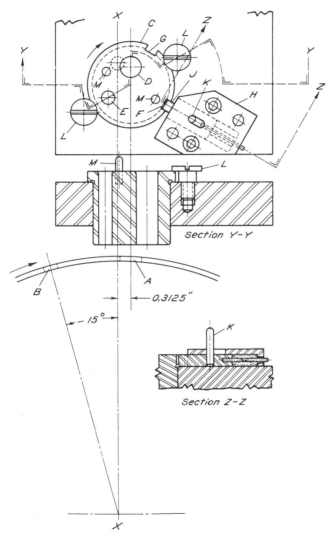

Fig. 4.   Illustrated position is to drill hole A. For drilling hole B, bushing hole E is on center line X-X.

During drilling, lock-screws L secure the bushing against movement. When the bushing is to be indexed, a pair of pins M provides a convenient means of rotating it.

Rapid-Indexing Drill Jig

The drilling of radial holes at various angular locations around the periphery of a cylindrical workpiece is an operation often required in the shop. The drill jig illustrated in Fig. 5 features rapid yet accurate indexing of such parts to speed their production. Although designed to hold workpieces having a central bore, other clamping arrangements can easily be adapted to the drill jig.

A baseplate made of aluminum supports two uprights spaced sufficiently apart to accommodate three 120-tooth gears mounted on a single shaft. Each of the gears is turned 1/3 of a tooth or 1 degree out of radial alignment with the other two gears. With this arrangement, the center of a tooth space is located at 1-degree intervals around the shaft. Bronze bushings in the uprights are line-reamed to provide a running fit for the gear-shaft. Another plate, positioned across the top of the uprights, has three drill bushings, each centered over one of the gears. A hardened pin is formed at one end in the shape of a gear tooth and the shank of the pin is made a slip fit in the drill bushings. The tooth-shaped end of this pin, seen enlarged at the right of Fig. 5 can be engaged with one of

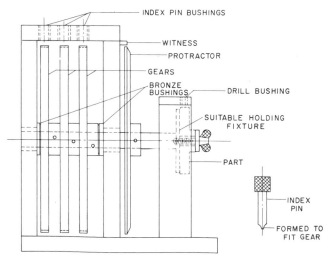

Fig. 5. Drill jig that permits accurate and rapid indexing for producing radial holes in cylindrical workpieces.

the gears to lock the shaft in a known angular position for the required drilling operation.

A plastic protractor and backing plate is mounted on the shaft in front of the forward upright. The protractor is graduated into 360 degrees in both directions. To facilitate use of the jig, each degree mark on the protractor should be identified in some manner with the bushing in which the pin is to be placed. Consecutive degree marks on the protractor of the original jig were painted blue, red, and yellow, and the colors were repeated in that order until all 360-degree marks were so indicated. The drill bushings were painted corresponding colors so that upon indexing to any particular angle both the mark on the protractor and the bushing in which the pin is inserted are identified by the same color.

The gear-shaft extends a length suitable to accommodate a holding fixture for the part to be drilled and a bushing plate, provided with an appropriate drill bushing, is supported over the holding fixture. With this arrangement, holes can be drilled at any angle, in full degrees, quickly and accurately. In addition, the jig can be used on end with the drill bushing plate mounted parallel to the diameter of the gears.

Multiple Drill Jig with Indexing V-Blocks

A novel feature of the multiple drill jig seen in Fig. 6 is the use of indexing V-blocks to locate workpieces of various sizes. Each of the six cylindrical V-blocks has four vee surfaces for holding workpieces 3/4, 7/8, 1, or 1 1/4 inches in diameter.

The jig was designed for drilling cross-holes near the ends of adjusting screws. Since the diameter of the cross-holes and their distance from the ends of the screws vary on different diameter screws, this jig obviates the need of individual tools for each size screw. Six screws of the same or various sizes can be drilled simultaneously, thus reducing the manufacturing cost per unit and also the lost or non-productive time required to locate and remove the parts.

A shaft passes through holes reamed in the seven spacer blocks that are secured to the bed of the jig. Collars are provided on the ends of the shaft, and a handle is mounted on the right-hand end for operating the jig. Attached to the bed, between the spacer blocks, are six baseplates, and secured to the sides of each baseplate are a right- and a left-hand side plate. These side plates are freely mounted on the shaft.

Suspended from a pivot-pin between each pair of side plates is a bushing plate. An oval-point socket-head set-screw is screwed through each of the six bushing plates, directly above the center of the shaft.

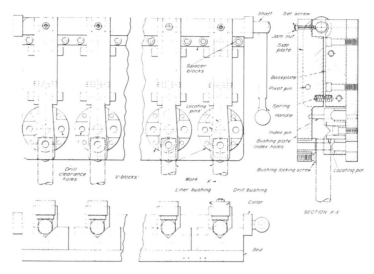

Fig. 6.   Multiple drill jig equipped with V-blocks, each having four vee surfaces, which can be indexed to accommodate workpieces of various diameters. The bushing plates are pivoted for loading and drilling by lowering and raising the handle, respectively.

The set-screws are held in vertical adjustment by means of jam nuts. Springs, held in pockets machined in the upper and lower surfaces of the baseplates and bushing plates, respectively, keep the oval points of the set-screws in contact with cam surfaces on the shaft. When the shaft is rotated by lifting the handle, the bushing plates are pivoted about their respective pins.

Driven into the forward end of each bushing plate is a liner bushing. Various drill bushings can be slipped into the liner bushings for drilling different sized screws. A locking screw, threaded into the under side of the bushing plate, holds each drill bushing in place.

The V-blocks, which are held to the base of the jig by means of hexagon, socket type cap-screws, are located radially, to align the proper vee surface with the particular size screw to be drilled, by index-pins. Each of these knurled-head pins passes through one of the four index-holes in the V-block and into a hole in the base of the jig. Four clearance holes are also provided in each V-block to allow the drill to break through the workpiece. Finally, four locating pins are driven into predetermined radial positions in the V-block, so that when the workpiece is abutted against one of the pins, the hole is drilled in the desired endwise location.

In operation, the spacing of the spindles on a Natco multiple-spindle drilling machine is adjusted to the center distances between the drill bushings on the jig. With the V-blocks indexed to their correct position, the handle on the jig is depressed. This allows the springs to pivot the bushing plates, lifting their forward ends so that the workpieces can be placed on the vee surfaces.

The outer ends of the adjusting screws to be drilled rest on a separate block (not shown) of the correct height. When the handle is lifted, the cam surfaces on the shaft pivot the bushing plates in the opposite direction, thus bringing the drill bushings into contact with the workpieces. Holes can then be drilled in all six parts simultaneously.

Template-located Jig Speeds Drilling and Counterboring

Frequently a set of holes has to be drilled and counterbored in a part with uneven spacing between the holes, which may be at different angles, and with odd spacing of the counterbores. This problem generally requires complicated tooling, and may slow up machining operations. To minimize these difficulties, the equipment shown in Fig. 7 was built, which consists essentially of two jigs.

In operation, the wedge-shaped work is first placed in jig No. 1, and drilled and reamed through bushing liner *B* on a drill press. Locating and clamping of the work is simple, and requires no description. The jig is then removed from the drill press and placed in jig No. 2 on a second drill press for drilling and counterboring the four holes *C* in the drilling and counterboring the four holes *C* in the required locations.

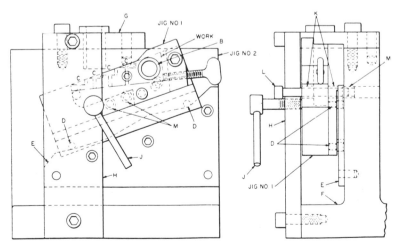

Fig. 7.   Two jigs used in combination with a template simplify the drilling and counter-boring of unevenly spaced holes.

Two pins D in jig No. 2 contact template E, which is doweled and screwed to the cast-iron fixture block F. A top plate G holds the principal liner for slip bushings. A bridge H is provided for lock-screw J. Jig No. 1 is brought into position for counterboring by sliding it along template E, so that index-pin L, passing through two axially aligned bushings K, enters the four bushings M in consecutive order.

Pins D, in contact with the template, together with the index-pin, provide three locating points that place jig No. 1 at the exact angle required for drilling and position it correctly for counterboring to the specified depth. When the index-pin is in position, lock-screw J is applied, after which the work is drilled and counterbored. To perform this operation at the next position, the screw is unlocked and the index-pin is moved from one bushing M to the next one.

Quick-acting Jig for Drilling Radial Holes in a Bushing

The drill jig illustrated in Fig. 8 provides a simple and rapid means of clamping work, insures accuracy, gives ample space for chip removal and burr clearance, and is easily unloaded. It is employed for drilling eight equally spaced radial holes near one end of a spring-cage bushing.

The jig body can be made from octagonal stock or from square stock with four flats machined on its periphery to form the necessary supporting surfaces for the jig when drilling the eight holes. One end of the body is bored to a diameter large enough to clear the head of the workpiece. The bottom of this hole is finished to provide a flat, square seat for the head end of the bushing to be drilled. A wide under-cut is also machined at the bottom of the hole to prevent chips from collecting

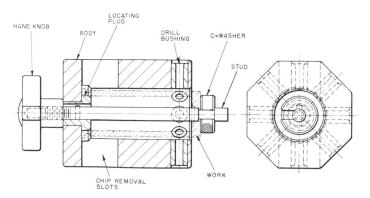

Fig. 8.   Simple, quick-acting jig employed for drilling eight equally spaced radial holes near one end of a spring-cage bushing.

under the end of the work. Two large slots are provided in the body, as shown, for the removal of chips from the hole.

Eight equally spaced radial holes are bored in the jig body, with their centers accurately located from the bottom of the work-clearance hole. Bushings for guiding the drills are pressed into the radial holes. This end of the body is made as short as possible, while still retaining sufficient stock to prevent the bushings from working loose. In this way, the work is allowed to project beyond the body to facilitate its removal from the jig after drilling.

A hardened and ground locating plug, the periphery of which provides a sliding fit for the bore of the workpiece, is pressed into the body at the bottom of the clearance hole. The nose of this short plug is chamfered to facilitate loading of the work. Projecting through, and keyed to the locating plug, is a long stud in which a recess is machined near the right-hand end to provide a seat for the knurled C-washer. A hand-knob is screwed on the locating plug.

After the C-washer has been slipped on the stud, the hand-knob is rotated until the work is firmly clamped between the bottom of the clearance hole and the inner face of the washer. At the completion of the drilling operation, the hand-knob is loosened, the washer is taken off, and the work is removed from the jig by pulling on its projecting end. If the part is difficult to remove, the stud can be pulled from the locating plug and pressure applied to the bottom of the workpiece by inserting a brass rod through the plug.

Adjustable Drill Jig for Unusual Shaped Castings

When a workpiece can be located in a jig or fixture from previously machined surfaces, positive contacts can be employed in locating. However, if the workpiece is a rough casting or forging, the locating surfaces or pins in the jig must be made adjustable to accommodate variations in the shape of the parts. An excellent example of an adjustable drill jig is seen in Fig. 9.

The workpieces, as indicated by broken lines, are right- and left-hand brackets for folding-top power units used on automobiles. The only machining operation required on these malleable castings is the drilling of a 25/64-inch diameter hole in each part.

Screwed into the left-hand end of the welded drill jig base *A* is a headless set-screw *B*. This screw can be adjusted to suit variations in the castings, and is locked in place by nut *C*. Secured in a slot in the rear wall of the jig base is a locating block *D*. A shouldered work-rest pin *E* is pressed into a vertical post on the jig base. Another vertical post on

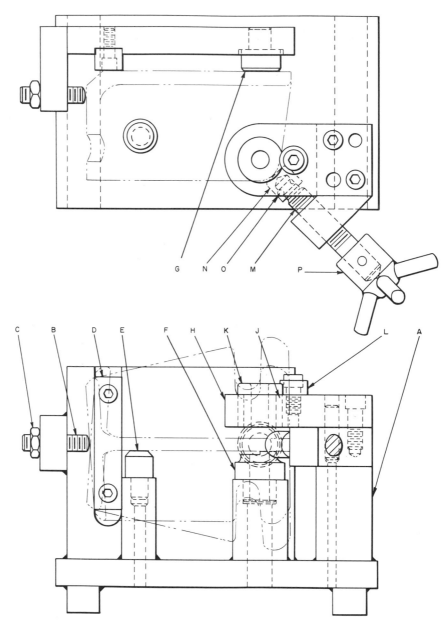

Fig. 9.   Jig employed for drilling both left- and right-hand bracket castings that have not been machined previously. Screw *B* can be adjusted to take care of variations in the shape of the castings.

the base holds a second rest-pin *F*. Positioning plug *G* is pressed into the rear wall of the base.

The bushing plate *H* is secured to the top of a vertical projection at the right-hand end of the jig base. Inserted in liner bushing *J* in this plate is a slip drill bushing *K*, which is prevented from rotating and being lifted with the drill by clamp *L*. Screw *M* is threaded through an angular projection on the jig base directly below the bushing plate. One end of this screw is reduced in diameter to accommodate a loose fitting, work-clamping plug *N*, which is prevented from dropping off the plug by a pin *O*. A four-pin hand-knob *P* is pinned to the opposite end of the screw.

With a workpiece resting on pins *E* and *F*, knob *P* is rotated to bring plug *N* in contact with the work. This movement forces the casting against set-screw *B*, block *D*, and plug *G*, and thus locates it for the drilling operation.

Jig for Burrless Drilling of Threaded Studs

When a cotter-pin hole is drilled in a stud with a rolled thread, a considerable burr is likely to be formed both where the drill breaks through and at the point of entry. If the compacted metal of a rolled thread is drilled, a larger and tougher burr develops than when the thread is die cut.

Deburring is, therefore, necessary, and is sometimes achieved by re-rolling the thread. This method is quick, but a portion of the burr may be forced back into the hole. Whereas the stud will then accept the nut, it may be necessary to clear the hole for assembly.

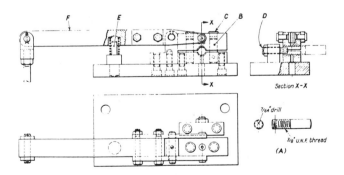

Fig. 10.   Jig for drilling cotter-pin holes in roll-threaded studs.

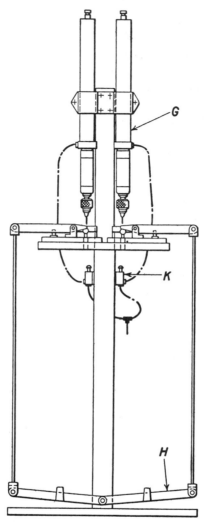

Fig. 11.   Dual setup for production drilling of the studs shown in Fig. 10.

A jig designed to overcome this difficulty is shown in Fig. 10. The stud to be drilled is gripped in the jig by two half-nuts. Consequently, the guide hole for the drill is continued to the bottom of the thread, and there is no space in which a burr can be formed.

The gripping halves of the jig are accurately aligned by posts B, of the die set type, sealed from chips by plugs C. A stop D controls the distance of the hole from the end of the stud. For ease of manufacture and accuracy, the clamping portions of the jig are made from one piece of tool steel. The locating hole is tapped and the guide-pin holes are drilled and reamed before the block is parted with a 1/16-inch saw.

For gripping the stud to be drilled, pressure is exerted by the spring E through the lever arm F. If this arm is controlled by a pedal, the operator's hands are left free to load the jig and operate the usual sensitive drilling machine feed.

A special-purpose arrangement comprised of two drill jigs and two air-operated, hydraulically controlled drilling units G is illustrated in Fig. 11. It is mounted on a 2- by 4-inch channel, welded at the lower end to a floor plate. A work-table supports the drill jigs, and the operator sits with a foot on each of two levers H.

In operation, one jig is loaded and a valve K for that unit is opened by the palm of the hand after loading. This action supplies air to the first drilling unit, and while that unit is in operation, the other jig is loaded and the valve depressed. Each unit retracts automatically when drilling has been completed, and remains retracted until valve K is again opened.

A smooth sequence of loading and drilling can be maintained. The hydraulically controlled feed and breakthrough permit drill sharpening to be carried out at regular intervals. Operator fatigue and drill breakage are kept to a minimum.

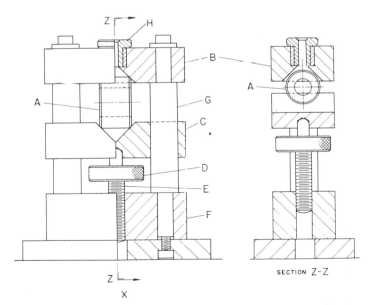

Fig. 12. A number of sizes of shaft collars can be drilled and tapped for set-screws in this drill jig. Fixed, upper jig plate B has a conical recess that centers each workpiece.

Drill Jig for Shaft Collars of Various Proportions

A drill jig that can hold any cylindrical component within a fixed range of outside diameters and widths is illustrated in Fig. 12. Parts are clamped on their outside edges between two plates — one having a conical recess, and the other, a V-groove. The tool was designed for drilling and tapping set-screw holes in several sizes of collars for use on steel shafts.

Instead of locating the workpiece by the bore, each collar A is positioned and clamped by means of its outer edges. To provide a better bearing surface, these edges should be given a 45-degree chamfer. A conical recess in the upper jig plate B centers each collar automatically. The adjustable lower jig plate C could have been provided with a similar conical recess, but the V-block shape was chosen since it resists the turning torque that is produced during the drilling and tapping operations.

In use, a collar is loaded into the jig by rolling it along the V-groove and simultaneously raising this lower jig plate to clamp the piece by turning the knurled knob D on the elevating screw E. The lower jig plate can be raised rapidly, as screw E has a coarse thread and turns freely in the tapped hole in the base F. A light clamping pressure is required to hold the work. Two supporting posts G for the upper jig plate also act as guides for the lower jig plate.

The half-section view X of the jig shows a large collar in place for drilling, while section Z-Z shows the jig with a small collar in position. As all collars required the same size set-screw, a single slip bushing H that is removed during tapping is used for guiding the drill.

This type of drill jig can also be adapted for drilling and tapping components such as spherical knobs.

Cross-drilling Two Shafts in One Setup

Drilling in-line holes through two cylindrical shafts requires a considerable amount of care, especially if the shafts are of different diameters. The device shown in Fig. 13 was used for this type of operation on bars ranging from 1/2 inch to 1 1/2 inches in diameter.

Referring to the illustration, shaft A, 5/8 inch in diameter, had to be drilled 0.125 inch diametrically, and shaft B, 1.0 inch in diameter, had to be drilled in-line to a diameter of 0.218 inch. The jig consists of two casehardened and ground V-blocks C and D, having 90-degree V-slots across one face. A drill bushing E is pressed into the upper V-Block. Baseplate F is machined to receive the lower V-block D, which is secured

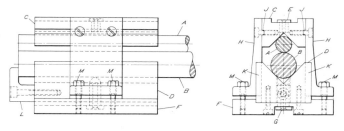

Fig. 13. Fixture with adjustable V-blocks for centralizing two shafts of different diameters for cross-drilling operations.

by cap-screws. Bushing *G*, pressed through the baseplate and into the V-block, provides for drilling the hole accurately in the larger shaft.

The V-blocks are aligned by means of two casehardened and ground L-shaped steel brackets *H*. Ground tongues *J* on the inner faces of the brackets fit tightly into machined grooves on the sides of the upper block. Two cap-screws on each side secure the V-block between the vertical inner side of the brackets. The sides of the larger block are slotted vertically at *K* to a sliding fit with the brackets. With the blocks and brackets connected in this way, correct alignment will be maintained.

To load the jig, the larger diameter shaft is placed in the vee of the lower stationary block and banked against a stop-bracket *L*. The smaller shaft is then placed in the jig with its end aligned with the end of the larger. Tightening the four clamping screws *M* equally will draw the upper V-block down evenly toward the stationary block. Both rods are tightly gripped and automatically centralized in correct axial alignment with each other.

In drilling the respective holes, the jig is first placed on the machine-table with the baseplate down so that the hole may be drilled in the small shaft. Then the jig is inverted for drilling the hole in the large shaft.

CHAPTER 10

Tooling for Grinding

Threading on a Tool and Cutter Grinder

A tool and cutter grinder doubles as an external thread grinder when equipped with a fixture which both revolves and advances the work in a helical path. Since the fixture is designed specifically for work of a given outside diameter and a given thread pitch, its use is more practical for quantity-lot grinding than for a single piece.

In Fig. 1, the work A is in the process of being ground by the wheel B. The grinding wheel is directly above the work, and the machine table is offset to the helix of the required thread. By turning handwheel C, a

Fig. 1.   By turning handwheel C, the spindle carrying workpiece A is both rotated and advanced beneath grinding wheel B.

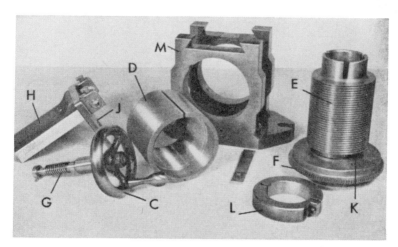

Fig. 2.   Since the pitch of nut *D* and spindle *E* must be the same as the thread required, the fixture is practical for quantity-lot grinding.

lead-screw spindle within stationary bronze nut *D* is made to revolve and advance toward the grinder wheel-head. The work is contained in the lead-screw spindle, the nose of which is a split collet.

In Fig. 2 are illustrated the disassembled parts of the fixture. The major body portion of spindle *E* is a lead-screw, identical in pitch (1/8 in this instance) to that of the ground thread required. This spindle engages bronze nut *D*, which is secured to frame *M*. A split in the wall of the nut permits it to be adjusted by set-screws to a snug fit with the spindle.

Worm-gear *F* is keyed to the rear end of the spindle, meshing with worm *G* to which handwheel *C* is attached. Dovetail slide *H* is a carrier for the worm, moving along the top of the frame as the spindle advances. This movement is transferred through the engagement of yoke *J* on the end of the slide with neck *K* located on the spindle adjacent to the worm-gear.

The spindle, a hollow member, has an inside diameter corresponding to the outside diameter of the work. To grip the work, collar *L* is tightened around the nose of the spindle as illustrated in Fig. 3.

Thread depth is controlled by the vertical adjustment of the grinder wheel-head. The rotation of the work through the worm-gear permits the needed slow feed, and also helps control the dwell in making a neat thread termination. The machine table also carries a thread-truing device. When the wheel needs to be trued, the table is unlocked and run over until the device is in position. A witness mark scribed on the table

Fig. 3.   The nose of the spindle is a split collet which grips the work *A* when collar *L* is tightened.

assures that it can be accurately returned to grinding position after the wheel is trued.

Fixture Facilitates Grinding of Spherical Radii

A spherical radius, to be machined on the end of a headed component, was required to be held to close limits of concentricity with its shank. To accomplish this, the fixture shown in Fig. 4 was designed. This fixture rotates the part and, at the same time, permits it to be pivoted while maintained at the desired radius setting.

Base *A* is of the turntable type, having a centrally located pin projecting upward from its top surface. Located on this pin, and screwed to the turntable, is a weldment *B* that serves as the frame for the fixture. Secured to the lower weldment member by cap-screws is a support block *C*. On this support block are mounted three other components of the fixture; an adjustable ball-end work stop *D* for maintaining proper radius setting, a group of four sealed ball bearings *E* for supporting the workpiece, and a handle *F* for swiveling the fixture.

A bushing *G*, through which a square hole has been broached (section Y-Y), is pressed into a horizontal support projecting from the main frame member. Square push-rod *H* rides within this hole. The push-rod is provided with a thin head at its upper end to retain compression spring J, and a hole tapped longitudinally in its lower end to receive the threaded shank of forked bracket K.

A hard rubber driving wheel L, which itself is driven through flexible shaft M by an electric motor, is supported by bracket K. This rubber wheel is offset at a slight angle, in a horizontal plane, so as to continually force the workpiece against work stop D. Adjustment of this angle is permitted by means of the threaded shank of the forked bracket together with lock-nut N.

The entire fixture assembly is mounted on the table of a standard cylindrical grinding machine. A workpiece is cradled between the ball bearings on the support block and banked against the ball end of the work stop as shown. Quick-acting clamp O is depressed, forcing the push-rod and driving wheel down, against spring pressure, until the rotating wheel contacts the workpiece shank. With the workpiece being thus forced to rotate, the spherical radius may be generated by swiveling the fixture with handle F. The reciprocating table movement of the grinding machine should be engaged.

Magnetically Operated Vise

When nonmagnetic components are to be ground and the production of special precision fixtures is not justified, it is normally necessary to hold each piece in a vise or by an adhesive. With either method, loading is slow, and it may be difficult to maintain the required degree of accu-

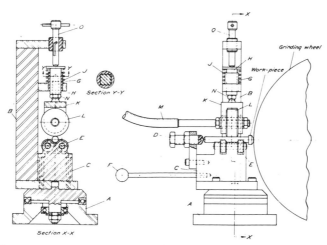

Fig. 4. Fixture rotates workpiece and also provides pivoting motion to simplify accurate grinding of spherical radii.

racy. Trouble may also be experienced when certain steel components are held magnetically, due to distortion under the clamping force and subsequent recovery.

To overcome these difficulties, the magnetically operated vise, shown in Fig. 5, was designed. This vise is simple and easy to produce, and where multiple loading is required, a number of them can be made economically to utilize the full area of the magnetic chuck. When made as illustrated the vise is suitable for holding small workpieces, but the dimensions can be varied as required.

In construction, a mild steel block, which forms the body, is milled to receive a clamping jaw *A* made of the same material. The body and jaw are drilled to take the pin on which the jaw pivots. There should be a diametral clearance of 0.010 inch between the pin and the hole in the jaw, and the latter should have a clearance of 1/16 inch on each side in

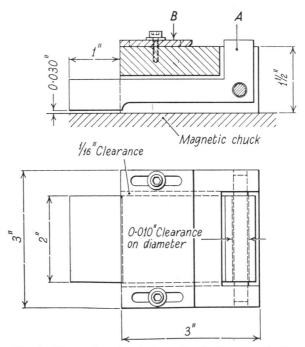

Fig. 5.   Plan and sectional views of a simple magnetic vise.

the slot in the body. The undersurface of the clamping jaw is relieved except at the left-hand end. An adjustable jaw B, which is made from a piece of plate, is secured to the body and set by means of two 10-32 UNF cap-screws. In use, a component is placed in the jaws and a 0.030-inch feeler is inserted between the toe of the clamping jaw and the magnetic chuck. Jaw B is slid forward to obtain the required width of opening and set. The feeler is then removed.

Clamping force can be varied, within limits, by increasing or reducing the gap between the toe of the clamp and the chuck surface. It is also possible to modify the force by changing the proportions of the jaw to obtain a different lever ratio. For general purposes, a ratio of 3 to 1 is satisfactory. Consistent results are obtained when the width of the workpieces is held to within limits of plus or minus 0.005 inch.

When the chuck is energized, the toe of the clamp is pulled downward, but remains clear of the surface. The pivoted arrangement assures that a downward thrust is exerted on the work to hold it against the location surface.

If narrow jaws are used, either singly or in multiples in one block, the depth of the toe portion should be increased to provide a mass adequate for the necessary magnetic attraction. With multiple jaws, the spacing must coincide with that of the magnetic chuck poles in order to retain full holding power.

As an indication of the results obtainable, hardened components 3 by 3/4 by 1/8 inch in size were required to be ground to remove 0.005 inch or more distortion caused by heat-treatment. It was necessary to grind one face of each piece flat before loading the parts in lots for finishing the opposite sides. With the magnetically operated vise, it was possible to obtain flatness within 0.0004 inch.

An advantage of this method of holding for such an operation is that the clamping is not "solid." Consequently, when the hardening "skin" is removed, each component can expand or resume its natural shape before the finishing cut is taken.

Gage for Adjusting Work-rest on Centerless Grinding Machine

One of the important factors to be considered in successful centerless grinding is the height setting of the work-rest. With many types of centerless grinding machines, this setting must be made by simple rule measurements. Ordinarily, these measurements are taken vertically from the upper edge of the work-rest to some stationary horizontal machine surface located above this site.

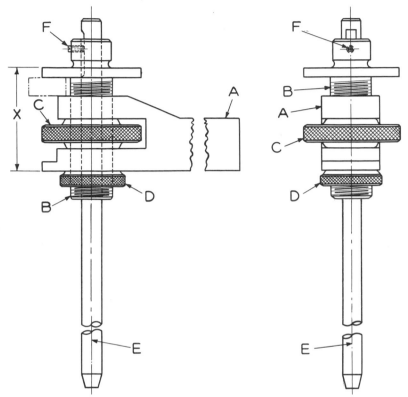

Fig. 6.   Adjustable gage simplifies setting of work-rest on a centerless grinding machine.

To expedite accurate work-rest adjustment, the gaging device shown in Fig. 6 will be found useful. Member *A* of the gage is a rectangular steel support bar that has been ground flat and parallel on all four sides. The length of the bar is determined by the size of the top surface of the regulating-wheel housing.

Fitting closely within a smooth bored hole in the forked, left-hand end of the support bar is the threaded shank of flanged sleeve *B*. Circular adjusting-nut *C*, which is a close fit within the accurately formed slot in the support bar, is bored and tapped to engage the fine-pitch threads on the shank of the sleeve. Rotation of this nut thus raises or lowers the sleeve within the bar. Knurled circular lock-nut *D* is threaded onto the lower end of the sleeve.

Sleeve *B* is bored concentrically to receive measuring rod *E*, which is a close fit therein. The rod is retained by means of set-screw *F*, which passes through the side wall of the sleeve head and bears against a flat

that is machined on the side of the rod. This provision allows the rod
to be adjusted to different positions within the sleeve.

A step, or ledge, is formed integrally on the lower forked end of support bar *A*, projecting sufficiently to bring it in vertical alignment with
the periphery of the flange on sleeve *B*. Micrometer measurements are
taken from the upper surface of the flange to the lower surface of the
projecting step; therefore, these two surfaces should be flat and parallel.

In application, support bar *A* is placed on the machined upper surface
of regulating-wheel housing *G*, Fig. 7. This surface, on most centerless
grinding machines, is convenient for gage location because of its proximity to the grinding wheel.

Measuring rod *E* should first be positioned within sleeve *B* so that a
simple micrometer reading is obtained between the sleeve flange and the
projecting ledge of the support bar (distance *X*) when the rod tip is on
the grinding wheel center line. This datum dimension should be used
as a constant from which to add or subtract when lowering or raising
the measuring rod with respect to the grinding wheel center line. Set-
screw *F* is then tightened so that the measuring rod and sleeve will move
in unison during the gaging operation.

When this preliminary adjustment of the gage has been completed,
nut *C* may be turned to move the measuring rod to the desired setting,
either above or below the center line of the wheel. If the work-rest is to
be located a certain distance above the wheel center line, this amount
should be added to distance *X*. On the other hand, if the work-rest is to

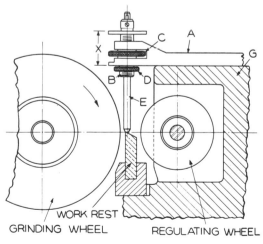

Fig. 7. Support bar *A* rests on machined top surface of housing *G* while measuring
rod *E* gages the work-rest setting.

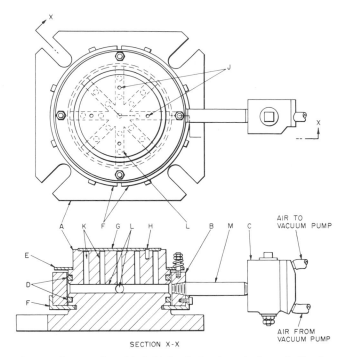

SECTION X-X

Fig. 8. Thin, non-ferrous disc *G* is held firmly in place during grinding by creating a vacuum in the chuck. When the setting of valve *C* is changed, the workpiece is blown from the chuck.

be located below the wheel center line, this amount should be subtracted from distance *X*.

In cases where a particular setting is repeated frequently, a suitable gage-block (shown in phantom lines, Fig. 6) can be made to fit between the upper surface of the support arm and the lower surface of the sleeve flange. Tightening lock-nut *D* will secure the setting.

The pre-set gage is placed on regulating-wheel housing *G*, as shown, with the end of the measuring rod lying immediately above one end of the work-rest. This member is then adjusted until its crest contacts the measuring rod tip. The gage is then slid across the machine housing until it lines up with the opposite end of the work-rest, following which that end is adjusted to contact the measuring-rod tip. In this way, the work-rest is set in a true horizontal position at the required distance from the center line of the grinding wheel. By using a different setting at each end of the work-rest, it can be set to produce any desired taper on the workpiece.

Vacuum Chuck for Holding and Ejecting Thin Discs

Thin, non-ferrous discs can be securely held during grinding and ejected at the completion of the operation by means of the vacuum chuck shown in Fig. 8. Since non-ferrous parts cannot be held by magnetic chucks, this device provides a convenient method of exposing the entire upper surface of the disc to the grinding wheel. With the air valve set in one position, a vacuum is created on the under side of the disc, which holds it firmly on the chuck by atmospheric pressure. When the valve is reversed, the ground disc is blown from the chuck.

Body *A* of the chuck is fastened to the rotating table of a vertical-spindle grinding machine, while the stationary ring *B*, valve *C*, and the electric motor and vacuum pump (not shown) are secured to the frame of the machine. O-rings *D*, washer *E*, and a split flanged washer *F* keep the chuck air-tight.

Workpiece *G* is placed in a nest on top of the chuck body, and a pin *H*, pressed into the body, enters one of the four holes *J* previously pierced in the disc. The chuck body is drilled to provide the vertical air passages *K* and the horizontal air passages *L*. Pipe *M* connects one of the latter passages to the air valve *C*.

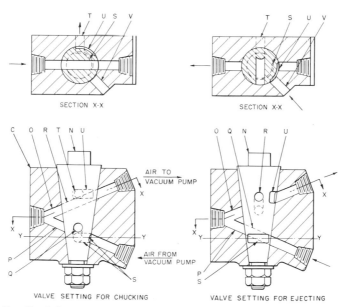

Fig. 9. Details of the special air valve employed on the vacuum chuck shown in Fig. 8. The valve setting for holding the work on the chuck is seen at the left, and that for ejecting the part at the right.

Enlarged views of the special air valve are shown in Fig. 9, with the valve setting for chucking seen at the left, and for ejecting at the right. Valve body *C* is bored to accommodate the tapered plug *N*. Dotted lines in both plan views show the cross-section of the plug in plane Y-Y.

Two angular air passages *O* and *P* are provided in the valve body, and two air passages *Q* and *R* in the tapered plug. While the angular holes in the body are in the same vertical plane, those in the plug are in vertical planes at 90 degrees to each other. Consequently, when the valve is set for chucking, as seen at the left, hole *R* is aligned with hole *O*, and air is sucked from the air passages in the chuck to create a vacuum under the workpiece. At the same time, hole *P* in the valve body is blocked off by the tapered plug. Thus air from the vacuum pump passes through the right half of hole *P* and an annular groove *S* on the plug, and finally, through a hole *T* to the atmosphere.

When the plug is rotated through an angle of 90 degrees to set the valve for ejection, hole *Q* is aligned with hole *P*, and air from the vacuum pump is blown into the air passages in the chuck, lifting the work from the chuck.

Evacuation of air from the chuck is prevented by the tapered plug sealing the left half of hole *O*. However, air from the atmosphere is drawn into the pump, through hole *V* in the body, around annular groove *U* on the plug, and out the right half of hole *O*. In this position of the air valve, hole *T* is sealed from the atmosphere by the tapered plug.

Fixture Facilitates Grinding of Eccentric Work

In the grinding of eccentric parts, problems are sometimes encountered by the manufacturer in maintaining close tolerances. Such a problem arose in planning to machine eccentric workpieces of the type shown in Fig. 10. The cylindrical and tapered end *a* is offset from the main portions of part *b* and flange *c*, which have a common center line.

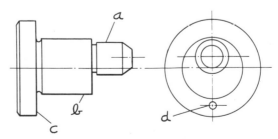

Fig. 10.   Eccentric workpiece conveniently ground by employing the fixture shown in Fig. 11.

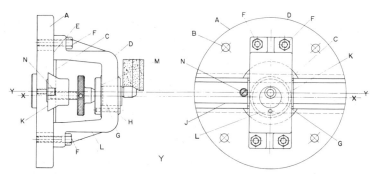

Fig. 11. Fixture designed to facilitate maintaining close tolerances in grinding of workpieces having an eccentric cylindrical surface.

The fixture in Fig. 11 was designed to locate diameter *a* concentric to the machine spindle and at the exact amount of offset required. Baseplate *A* has a flange turned on the back face for mounting on the machine spindle and is secured by four cap-screws through *B*. Shoulder *C* is turned on the front face of the baseplate to locate the fixture concentrically. Member *D* of the fixture is a cast-iron bridge with the top face *E* bored to fit on shoulder *C*. Four cap-screws *F* hold the bridge to the baseplate.

Hole *G* is bored through the bottom face of the bridge (at the suitably offset centerline X-X) to receive the part, which is located in the fixture by dowel-pin *H*. This pin engages closely in hole *d*, previously machined in the workpiece. Care must be taken to bore hole *G* in the fixture and the dowel hole in the work with their center lines offset the required distance from centerline Y-Y to insure accurate positioning of the work.

To clamp the workpiece in the fixture, a dovetailed guide way *J* is machined diametrically across the front of the baseplate to fit slide *K*. This slide carries the clamping screw *L*, which has a hardened nose that presses against the back of the work and holds it firmly against machined face *M* on the bridge. The nose of screw *L* should bear approximately against the center of shoulder *c* on the workpiece. A stop-pin *N* is added to regulate the position of slide *K*.

Simple Device Speeds Taper Setups for Cylindrical Grinder

A 10-inch sine bar, plus a center-mounted support platen, enables precision taper settings to be easily and quickly made on a cylindrical grinder. The support platen is a simple bar of 3/4- by 4- by 18-inch flat, cold-drawn steel, which has been carefully surface-ground all over and center holes drilled in the ends exactly on its center line. The sine bar

Fig. 12. With platen on surface plate, sine bar is clamped at a desired taper angle using a vernier height gage.

is clamped on the platen and set at the desired angle Fig. 12, using a surface plate. For most work a vernier height gage is sufficiently accurate. However, gage-blocks can also be conveniently used for setups requiring a higher degree of accuracy.

After the angle has been set, the platen is seated on the centers of the grinder, positioned, and clamped parallel to the table (across its width)

Fig. 13. Platen is set parallel with table, and held on centers. With dial indicator clamped to wheel, a continuous zero reading shows taper is same as sine bar's original angle.

by use of the solid and adjustable parallels, as shown in Fig. 13. A dial indicator, clamped in any convenient manner to the wheel or wheel-head, can be used to verify the angular setting of the grinder table. A zero reading throughout the length of the sine bar indicates that the machine is set to the same angle that was established for the sine bar on the surface plate.

Grinding Fixture that Insures Desired Work Height

Some types of work, particularly parts having an angular or irregular base, must be held in special fixtures while being finished on a surface grinding machine. On jobs of this category it is often difficult to measure the height of the completed piece, and so the finish cut becomes a tedious cut-and-try process.

A fixture designed to facilitate operations of this kind is illustrated in Fig. 14. This fixture is provided with a built-in adjustable diamond for truing the working surface of the wheel exactly on the plane of the desired work-surface. It conditions the wheel for a high-quality finish.

The operator simply traverses the grinding wheel across the diamond after he has made a vertical adjustment of the diamond to within a few thousandths or possibly ten-thousandths of an inch of the sizing cut. With this truing arrangement each successive piece of work will be identical regardless of wheel wear or other variables which might enter into individual setups of a production job.

Bore-grinding Fixture for Bevel Gears

Grinding the bore and face of a bevel gear accurately concentric with the teeth, after heat-treatment, often presents difficulties when the only possible location is from the tooth surfaces. Various types of fixtures

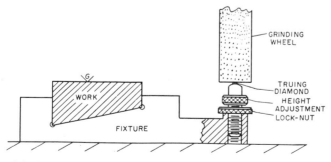

Fig. 14. Grinding wheel truing device mounted on work-fixture to insure accurate work height.

have been designed for this purpose. Many of them, while fairly satis-factory in operation, are expensive to produce, and easily lose their initial accuracy through wear of small parts, such as balls or pins, which are used to locate the work from the gear teeth. After some experience with the more complicated types of fixture, the design shown in Fig. 15 was developed, which not only proved quite successful, but also had the advantage of being comparatively inexpensive.

The gear to be ground, shown at *A*, is located in slots in the ring *B*, the number of slots corresponding to the number of teeth in the gear. The locating ring is fitted in the body *C* so that it is easily replaceable, and can also be reground on the front face while in position, should any wear or damage occur at the edges of the slots. Since the slots are parallel to the axis, grinding of the front face of the ring does not result in any change in their form.

The work is clamped by a ring nut *D*, the threads of which are made an easy or slack fit on the body. As an alternative to the ring nut, three clamps could be used, operated either independently or by a draw-bar with an equalizing device. The ring nut shown, however, is satisfactory, providing the threads are a slack fit, as stated, so that the back face of the gear controls the position of the nut.

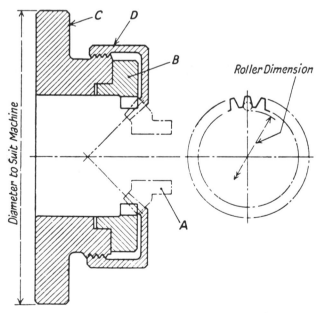

Fig. 15.   Fixture for holding bevel gear to permit accurate grinding of bore concentric with pitch circle.

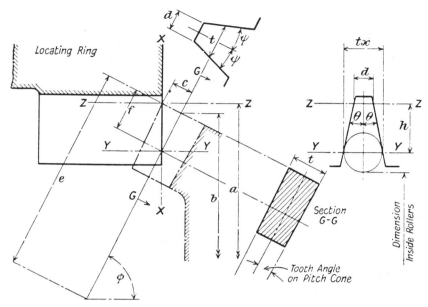

Fig. 16. Diagrams illustrating method of making enlarged scale layout drawings for determining shape and size of tooth-locating slots in fixture shown in Fig. 15.

Referring to Fig. 16 which shows the dimensions determining the shape and size of the work-locating slots:

ψ = pressure angle of gear on back face;

θ = angle of slot in locating ring;

ϕ = pitch cone angle of gear;

a = outside diameter of gear;

b = pitch circle diameter of gear;

c = addendum of gear tooth;

d = thickness of tip of basic rack on outside diameter a, and slot width on the line ZZ;

e = cone distance of gear;

f = distance from back face of gear to face XX of locating ring;

t = circular thickness of gear tooth on pitch circle diameter b;

t_x = width of slot on datum line YY;

h = distance from lines ZZ to YY;

YY = datum line through pitch cone line, and intersection of face of fixture XX;

ZZ = datum line at right angles to outside diameter a;

XX = front face of locating ring.

The values of d and the angle θ can be found either by making an enlarged lay-out as shown, or by calculation, using the following method:

Given the values t, c, and ψ from the gear drawing, then

$$d = t - 2\,(c \tan \psi)$$

The thickness of the tooth t_x on the line YY, and also the distance h, must next be determined in order to find the value of θ.

$$h = c \sec \phi \text{ and } f = c \cot \phi$$

then

$$t_x = \frac{e - f}{e}\,t$$

and

$$\tan \theta = \frac{(t_x - d)}{2h}$$

Having obtained the above values, it is a simple matter to calculate an inside roller dimension for use when making the locating ring. The slots are cut slightly deeper than the outside diameter of the gear to provide clearance, while the bore of the ring is such that it is well clear of the root line of the bevel gear teeth so that the gear can be held in place as shown in Fig. 15.

Grinding Wheel Radius Truing Device

Occasionally it is necessary to dress a grinding wheel to a large radius in order to obtain some desired curvature in a forming cutter or die part. The device shown in Fig. 17, designed specifically for a tool and cutter grinding machine, can be used to produce either a convex or a concave shape on the periphery of a grinding wheel during truing.

Directly beneath the wheel A and extending over the rear of the machine table B are two rectangular steel plates C and D. The front end of the lower plate is held on the magnetic chuck E of the grinder. A cable rig F supports the rear end of this plate. Both plates are provided with a series of holes G accurately located on 1-inch centers. Holes in the upper plate are 1/16 inch larger in diameter than those in the lower plate, in order that the fulcrum bolt H can be seated properly.

The diamond dressing tool J is supported on the center of the wheel by a post K. This post is so fastened to the upper plate that distance a is exactly 1 inch. To adjust the tool, gage piece L is temporarily attached to the post. A shelf on the gage piece, 1.250 inches wide, accommodates one or more size blocks. In the illustration, only size block M is required.

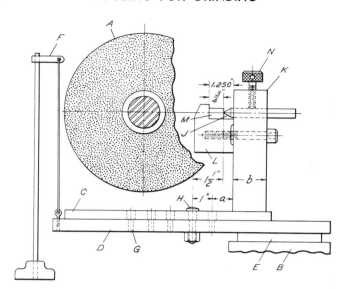

Fig. 17.   Plate *C* pivots around fulcrum bolt *H* to produce the desired radius in the periphery of wheel *A*.

If, for example, a 1 1/2-inch convex radius is required, the fulcrum bolt occupies the second pair of plate holes, as illustrated, and the diamond tool is set against a 3/4-inch size block. Thumb-screw *N* locks the tool in position, after which the gage piece is removed. With the wheel revolving, plate *C* is pivoted manually, and the table saddle is slowly advanced toward the grinding wheel.

By constructing the post so that width *b* is 1.000 inch, the device can be used to produce a concave wheel periphery. In this instance, the plates are reversed so that they extend over the front of the table, the gage piece is attached to the opposite side of the post, and the diamond tool is reversed in the post.

The wheel of a conventional surface grinder can also be dressed with this device. Since the wheel axis is at right angles to the stroke of the table on such a machine, the plates can be held without the need for a supporting cable and bracket rig *F*.

Simple Hold-down Fixture used for Truing Diamond Wheels

Diamond cup grinding wheels are often reconditioned by lapping the working face on a flat plate with loose abrasive grains. After this operation has been completed, it is necessary to bring the back face of the diamond wheel into parallelism with the reworked face in order that the

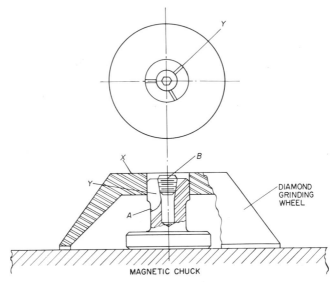

Fig. 18. Simple fixture applied for holding diamond grinding wheels on a magnetic chuck in truing operations.

wheel will run true. Rather than going through the tedious process of shimming this back face of the diamond wheel against a shoulder on the spindle adapter, the writer simply surface-grinds the wheel seating surface by means of the machine setup shown in Fig. 18.

Use is made of a precise hold-down fixture A which is slipped into the arbor hole of the grinding wheel. The fixture and wheel are then seated on the magnetic chuck of a surface grinder. It is essential that the chuck surface be true with the axis of the machine spindle. When both the wheel face and the base of the fixture are seated on the chuck the taper pipe plug B is screwed down into the split upper end of the fixture. This secures the fixture to the grinding wheel. The upper end of the fixture has been split by milling three equally spaced slots Y through it. A light cleanup cut on surface X on the diamond wheel will readily bring this surface parallel with the face surface within 0.0002 inch.

Fixture for Use in Grinding Angles and Rounded Corners

Various combinations of straight and curved surfaces can be finish-ground by using the general-purpose fixture shown in Figs. 19 and 20. Rapid set-up and simple operation are two desirable features of this device.

Fig. 19. Typical workpiece, having tapered ends and rounded corners, that can be ground to shape on the general-purpose fixture illustrated in Fig. 20.

A workpiece that requires a rounded corner having a radius of 0.812 inch on each end and angular surfaces of 7 degrees is illustrated in Fig. 19 by the solid lines. The blank from which this part was made is shown by dotted lines. One rounded corner and tapered side are scribed on the face of the blank. The work is then turned over and the same procedure is followed on its opposite end. Locations of the radii should be center-punched. Excess stock is next removed on a band saw to facilitate grinding.

Top plate A and bottom plate B are attached to rectangular spacers C and D by cap-screws. These members form the boxlike base of the fixture, as shown at X in Fig. 20. Swivel-table E pivots about tapered stud F, a nut G on the stud permitting the table to be locked in any desired position.

Two table-stops, shown at Y, are used — stop H is pressed into slide J and stop K is pressed into slide L. Each slide is located against an edge of top plate A by two screws. The screws pass through long slots machined in the slides, thus providing a means of adjustment.

In setting up the fixture, it is placed on a magnetic chuck on the table of a horizontal grinding machine. The back edge of top plate A is indi-

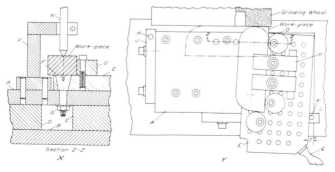

Fig. 20. Fixture which facilitates the grinding of workpieces having various combinations of straight and curved surfaces. Eccentric washers serve as readily adjustable banking points.

cated to bring it parallel to the path of the work-table. The workpiece is positioned on swivel-table E by a locating gage V which is doweled to top plate A, as shown at X. Sliding center N should be in direct alignment with the center of tapered stud F. The sliding center is used to align the punch mark on the face of the workpiece with the point about which the table will pivot.

Three eccentrically drilled washers O bear against the workpiece and serve as banking points. Straps P are used to clamp the work to the table. Locating gage V is then removed.

A protractor is employed to set table E at an angle of 7 degrees, as shown at Y. Table-stop K is then set against the side of the swivel-table and slide L is locked in place. Table-stop H, together with slide J, is adjusted so that the forward edge of the swivel-table will be parallel to the face of the abrasive wheel at the end of the grinding stroke.

To grind the 7-degree taper, table E is held against stop K by means of handle Q and locked by nut G. The machine table is then fed past the abrasive wheel, and the rounded corner is formed by pivoting the table until it contacts stop H.

After these surfaces have been ground on all the workpieces, they are turned over and the same procedure is followed in grinding the opposite tapered end and rounded corner. Locating gage V is used to line up the second punch mark, and the eccentric washer at the end of the part is adjusted accordingly.

Grinding Fixture for Holding Work at Any Angle

Workpieces can be held at any required angle by means of the grinding fixture shown in Fig. 21. The fixture, which can be mounted on a magnetic chuck or clamped to the table of a surface grinding machine, is simple to construct and use, but rugged and accurate.

The T-shaped body A of the fixture can be made from a well-seasoned piece of close-grained cast iron, drilled to provide three horizontal rows of five equally spaced holes each. Three screws B, having eccentric work-holding heads, are held in the body by washers and nuts.

By inserting the screws in the proper holes and tightening them when their eccentric heads have been rotated to the required positions, the workpiece C can be held at the necessary angle. Square and vernier protractors can be used to determine the location and adjustment of the screws for various angles. Once the screws have been set, any number of identical pieces can be ground without altering the setup or removing the fixture from the machine.

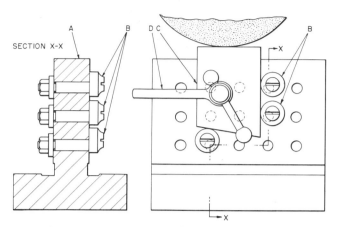

Fig. 21.   Versatile grinding fixture in which workpiece *C* is held on and against three adjustable screws *B*. Clamp *D* holds the work against the face of fixture body *A*.

A clamp *D* can be used to hold the work against one face of the fixture body. Figure 22 shows four of the many setups possible with this versatile fixture. Slight variations in the spacing of the holes will not affect the accuracy of the fixture, since such discrepancies can be compensated for by proper setting of the screws.

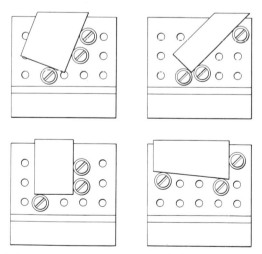

Fig. 22.   Four of the many setups possible on the versatile grinding fixture shown in Fig. 21. By inserting the screws in the proper holes and adjusting their eccentric heads, workpieces having various angular surfaces can be accommodated.

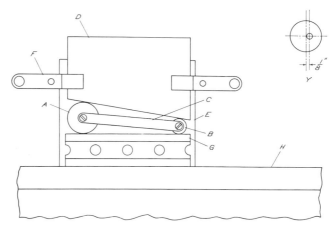

Fig. 23.   Work-rest that locates part from an angular bearing surface grinding.

Work-Rest for Surface Grinding Parts Seated on Angular Face

A work-rest, useful in locating work from an angular face for the grinding of flat surfaces, is illustrated in Fig. 23. Basically, the unit consists of two discs *A* and *B* attached to opposite ends of bar *C*. Workpiece *D* is held on vertical face of angleplate *E* by parallel clamps *F*. The work-rest assembly is supported by a parallel *G* located on the magnetic chuck *H*.

The center-to-center distance of the discs should be sufficient to enable the workpiece *D* to be seated firmly. The discs are machined to convenient diameters, provided the difference in the diameters in combination with their center-to-center distance gives the desired angle.

To find the difference in the diameters of the two discs, determine the sine of one-half of the desired angle (using a sine table) and multiply this value by twice the center-to-center distance of the discs. The discs should differ in diameter by this amount.

Provision is made for a slight adjustment of the angle by drilling and tapping disc *A* 1/8 inch off center, as indicated in view Y. The work-rest may thus be accurately set to a sine bar or a vernier bevel protractor. Once the desired angle is obtained, disc *A* is tightened firmly to bar *C*.

Work-holding Fixture for Rotary Surface Grinding

In machining a cast-iron gear housing for farm equipment, the first operation consisted of finishing a large flange. Accuracy and a good finish on the flange face were important not only because of product requirements, but also because all tool settings in subsequent operations had to be relative to this face.

If the operation were performed on a lathe, roughing and finishing cuts would be necessary, and so it was decided to machine the part on a rotary surface grinder. This necessitated the design of a work-holding fixture which should be of simple construction so as to permit quick amortization.

The unique fixture shown in Fig. 24 was designed to meet requirements. It consists primarily of base *A*, which is a piece of hot-rolled steel machined flat on both sides. The bottom surface was ground to insure good contact with the magnetic chuck on the grinder. Two uprights *B* of cold-rolled steel spaced 120 degrees apart were screwed and doweled to base *A*. On top of the uprights are two hardened and ground locators *C*. These locators are ground to a chisel point where they contact the work so as to hold the part rigidly while the grinding cut is being taken.

A third upright *D* was also attached to base *A*. This upright was machined with an angular surface to match a corresponding surface on clamp *E*. In setting up a housing for an operation, tightening of the hexagon nut *F* causes clamp *E* to dig into the work. At the same time the work is drawn down against three hardened and ground rest pads *G*. With one clamping motion the part is gripped inward and downward without danger of slipping.

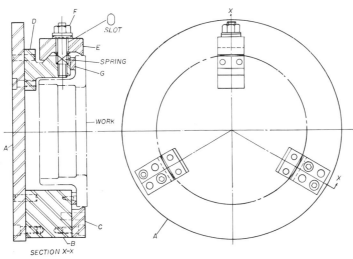

Fig. 24. Ingenious fixture devised for a flange-finishing operation performed on a rotary surface grinder. Chisel clamp *E* locks work when nut *F* is run up.

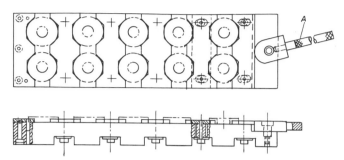

Fig. 25. Fixture designed for holding bronze washers in a production grinding operation.

Fixture for Production Grinding of Bronze Washers

A unique fixture designed for the simultaneous grinding of ten bronze washers is shown in Fig. 25. The washers are 1 1/2 inches in diameter and are ground on both sides to a thickness of 0.175 inch within plus or minus 0.001 inch. Both sides of the washers are chamfered so that there are no burrs to contend with. It is intended that the fixture be held on a magnetic chuck.

The washers to be ground are clamped between double V-blocks which were made from 5/32-inch thick flat ground stock, being, of course, slightly less in height then the thickness of the washers. Each V-block is attached to the base by means of two machine screws, spacing bushings, and small washers. These units are a slip fit in elongated holes that have been provided in the base. The elongated holes permit all of the V-blocks, with the exception of the one at the extreme left-hand end, to be moved lengthwise when clamping handle A is tightened against the right-hand V-block. In this way all ten bronze washers can be instantaneously clamped. Small variations in their diameters are permissible because of the elongated-hole arrangement.

Bushing Mandrel for Cylindrical Grinding

Conventionally, when finishing the periphery of hardened steel bushings on a cylindrical grinding machine, the workpieces are forced on a tapered mandrel with an arbor press. This process, plus the necessary manipulation of the set-screw on the driving dog, is, at best, time-consuming. When the production of a number of such bushings is required, quick and easy mounting and removal of the workpiece is accomplished

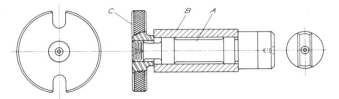

Fig. 26.   Mandrel for production of bushings on a cylindrical grinding machine.

by the use of the mandrel illustrated in Fig. 26.   In addition, the need for a driving dog is eliminated with this device.

The seating surface of the special mandrel A is ground to a snug fit in the bore of the workpiece B.   This bushing can then be either wrung onto the mandrel if the bore size is at the high-tolerance limit or lightly tapped on if the bore size is at the low-tolerance limit.   By relieving the center of the mandrel, accurate seating is obtained and resistance to the mounting of the work is reduced.   One end of the mandrel is threaded to receive the driving plate C.   This slotted, disc-shaped member not only engages with the head-spindle driving pin, but also acts as a nut to secure the work against slippage.   The knurled peripheral surface provides an effective hand-grip for loosening and removing the driving plate when required.

A tang, milled on the tail-spindle end, is used, if necessary, as a key for holding the mandrel.   When removing tight workpieces, the tang may be slipped either between the jaws of a vise or in a slotted block mounted directly on the work-table of the machine.

Fig. 27.   Components of the tubing grinder mandrel, reading clockwise from the left: arbor, rubber bushing, steel compression washer, and knurled take-up nut.

Fig. 28. Aluminum sleeve on the expansion mandrel which is mounted on centers of the grinding machine ready for precision grinding of the outside diameter.

Thin-Wall Aluminum Sleeves Ground Without Distortion

Because of the tendency of thin-walled aluminum tubing to distort when held for grinding on a mandrel, a special device, Figs. 27 and 28 was successfully designed and built, permitting the specified 0.001-inch roundness tolerance to be met.

Other requirements for this tool were that the work must be held securely; that it could be quickly and easily loaded and removed from the machine centers. A variety of different methods of holding the sleeves had been tried, but neither the standard type of expanding mandrel nor a simple rubber bushing mounted on a standard lathe mandrel was satisfactory in production.

The most practical mandrel incorporates a steel body and a semihard rubber bushing which is expanded by the knurled hand nut, forcing the bushing against a shoulder on the steel body, Fig. 27. The bushing was made from a cylinder of the type of rubber used to make a relatively firm press stripper. It was drilled and reamed to a tight fit on its special mandrel. Then (in assembly) it was turned and cylindrically ground to a slip fit with the aluminum tube stock. When a work sleeve is mounted on the mandrel it takes but the slightest movement of the knurled take-up nut to expand the rubber bushing enough to lock the sleeve for grinding. Releasing the nut allows the work to be stripped off the bushing without great effort. Uniformity of roundness and wall thickness throughout the length and circumference of the work is well within 0.001-in.

CHAPTER II

Gaging Devices

Induction Gage Measures Shaft Diameter During Grinding

Continuous measurement of the diameter of a shaft being ground is possible with the gage shown diagrammatically in Fig. 1. The device is a Russian development. It has no moving parts or electrical contacts, and works on the principle of a transformer with variable air-gap inductance.

Anvils A of the gage rest against shaft B being ground. The included angle of the anvils with the work, 2ψ, is 30 to 40 degrees. Inductance pickup C, built into the body of the gage, consists of two windings (supply circuit S and measuring circuit M) and a magnetic core which is open on one side.

As the diameter of the shaft decreases, the distance from its surface to the core of the pickup also decreases. For a decrease in shaft diameter from D to d, the distance changes correspondingly from L to l. From the geometry of the diagram,

$$L - l = \frac{d}{2} + \frac{D - d}{2 \sin \psi} - \frac{D}{2}$$

or

$$L - l = \left(\frac{D - d}{2}\right)\left(\frac{1 - \sin \psi}{\sin \psi}\right)$$

thus if $2\psi = 30$ degrees,

$$L - l = 1.44 \, (D - d)$$

and if $2\psi = 40$ degrees,

$$L - l = 0.96 \, (D - d)$$

Voltage induced in the secondary winding (measuring circuit M) is a function of the size of the air gap between the work and the core.

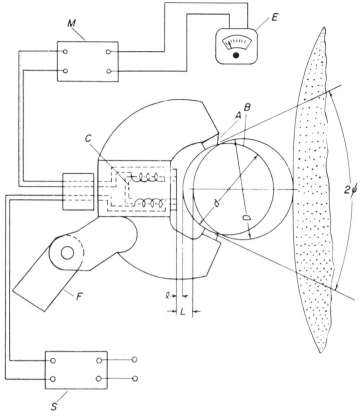

Fig. 1. As the shaft is reduced in diameter, the distance between its surface and the core of the pickup decreases, causing a change in the voltage of the measuring circuit.

The measuring circuit M is connected to a scale-reading instrument E having 150 divisions. Each division represents 0.2 micron (0.000007 inch) or more, as required. When grinding has reduced the shaft diameter to proper size, the wheel is retracted either manually or automatically. In the latter instance, a relay is incorporated in the measuring circuit.

The anvils are made of tungsten carbide. Since the pressure on each is only 150 grams, wear is insignificant. Arm F is fixed on the grinding machine table, and provides a swivel support for the gage.

If several diameters have to be ground, a gage can be provided for each one, the pickups in each being independently adjustable. Setting can be performed with two master workpieces temporarily held between the machine centers, one representing the upper limits of the tolerances, and the other, the lower. All pickups are, as a rule, adjusted to give the same reading for the lower-limits master. The air gap should not be more than 0.0006 inch for the lower limits.

Sine Bar for Use with a Combination Square

The machinist's combination square equipped with a protractor head is an extremely versatile and useful instrument, but the graduations on the angular scale are not fine enough for precise work. The sine bar illustrated in Fig. 2 will, if carefully made, permit setting the instrument to an angle accurate within a few minutes of a degree.

Simple to make and easy to attach, this sine bar is convenient to use when a surface plate or other precision flat surface is available. A piece of flat bar stock is machined as shown in Fig. 2 to form the body A. The material is mild steel and no heat-treatment is necessary. There is only one highly critical measurement involved in the construction of the attachment. The two standard hardened and ground dowel-pins B must be spaced an exact distance apart, center to center. These dowels are 3/16 inch in diameter and are made a press fit in reamed holes.

Normally, the sine bar will be used only for setting scale C of the combination square to the desired angle. Therefore it will not be necessary to machine the top and bottom edges of the bar to exact parallelism with the dowel-pins, unless these edges are to be used as gaging surfaces. Dowel-pins B serve two purposes. First, they act as buttons from which all measurements and settings are determined. In addition, by extending through body A, these pins insure alignment of the edge of the blade with the buttons when the bar is clamped in place for use.

Two knurled thumb-screws D of any convenient size are provided. A 10-24 thread is suggested. When tightened, these screws pull the dowel-pins tightly against the lower edge of the blade and hold the sine

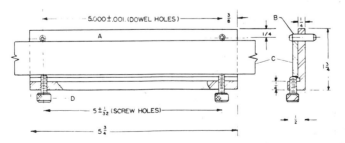

Fig. 2.   Sine bar for accurately setting the blade angle of a combination square that is mounted in the protractor head.

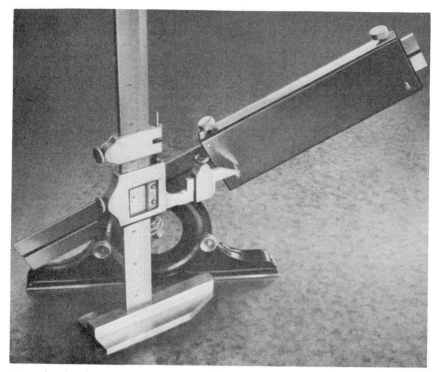

Fig. 3. A vernier height gage is used to obtain the vertical distance between the centers
of two pins.

bar in place while the angular setting is made. For convenience, this particular sine bar has been designed with a button center-to-center distance of precisely 5 inches.

In use, a preliminary setting for the angle of the scale can be quickly made from the graduations on the protractor head. Then final adjustment is obtained by use of the sine bar. A vernier height gage is employed to set the angle of the blade precisely by measuring the vertical distance between the buttons on the bar (Fig. 3). This arrangement also serves as a double check against possible error.

Novel Gage Checks Internal Grooves

Capable of exceptionally fine gaging, a device (see Fig. 4) whose principle can be used for inspecting outer ball-bearing raceway grooves and a variety of other internal bores has been built for use in a valve-making plant. The raceways are 4 inches and over in diameter and are components of large gate valves. Different diameters having the same tolerance can be set up by changing the adjustable pin on the left in the sketch. The tolerance can be changed by shifting the dial center hole a calculated distance to the right of its true center.

The gage consists of four machined parts: a base *A*, a dial *C*, and two pins *D* and *E*. The dial *C* (see Fig. 4) is bored 0.002-inch off-center for the fillister-head screw *B*. The eccentric dimension can be a value equal to or less than the full tolerance of the work part. The dial itself is calibrated after assembly, using a micrometer. Thus, the eccentric dimension is not critical. The gage in the sketch is calibrated for a 0.004-inch tolerance. In practice, simple minimum and maximum marks are usually satisfactory for production inspection purposes.

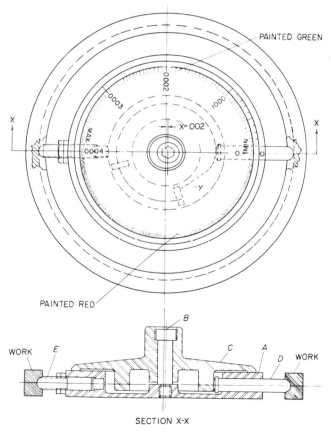

Fig. 4. Simple gage for bore grooves, such as ball-bearing outer races, gives fine measurements that are easy to read quickly.

The illustration shows how large rotary movements of the dial C act in the manner of a cam on the sliding pin, which acts as a follower. The pin is pushed outward in extremely small proportional amounts. As an aid in centering the work in the device, there is ample room to add spring-loaded pins at 30 degrees on either side of the adjustable pin E.

This same operating principle can be applied to depth-gaging devices, with the advantage that they are much easier to read than regular flush-pin gages.

Gage Checks Concentricity of Threaded and Reamed Holes

An interesting gaging setup was designed to check the concentricity between a threaded hole in an outside wall and a reamed hole in an inside wall. Since the hole being checked was partially hidden, one of the major considerations was that the gage be read from the outside. The unit that was designed for this particular job is shown in Fig. 5. Construction cost was small, as was the time required for its use.

Basically, the gage body consists of two bushings. Outer bushing A is made from SAE 3150 steel that has been hardened to Rockwell C, 42 to 50. It has a male thread accurately ground to fit the female thread in outer wall B of the workpiece, in this case a 1 1/2-18 NEF-3 thread. A central hole in the bushing is ground to a slip fit with inner bushing C. This bushing, made of hardened and ground tool steel, contains a through hole and a milled slot to receive a standard dial-indicator hole attachment D, which is secured with headless set-screw E. Spring plunger F maintains constant pressure against stem G of the hole attachment to keep it in contact with the stem of a standard, flat-back dial indicator H.

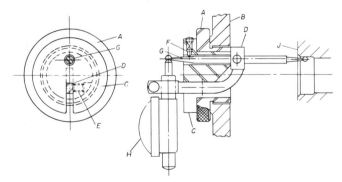

Fig. 5.   Concentricity of threaded hole in wall B and reamed hole in wall J can be quickly checked with this simple gage. With bushing A threaded into the outer wall, bushing C is rotated and eccentricity is noted on dial indicator H.

In use, bushing *A* is threaded into the workpiece by hand. As this is done, the end of stem *G* enters the reamed hole in inner wall *J*. Bushing *C* is then rotated, and the eccentricity of the reamed hole can be obtained by direct reading of the dial indicator.

Tooling Ball with Removable Feature

Tooling balls enjoy high popularity as reference points of measurement. These balls are accurate and versatile and are often preferred to the use of a machined surface, dowel-pin, or tooling button. The conventional tooling ball, however, does have certain shortcomings.

First of all, the ball creates an obstruction, and its surface is exposed to damage or alteration. Secondly, since the ball is a permanent part of the tool, it becomes an idle investment for the most part, especially when the tool is in storage.

In the third place, the ball cannot always be quickly, accurately, and permanently set. Often, a great deal of time is consumed in tapping it into and out of its base pad to obtain the desired height setting. Because the setting is always susceptible to change from an accidental bump, a constant check must be made of its height above the pad.

Finally, the stem of the conventional tooling ball offers interference when certain angular readings are attempted, as will be explained. It is also desirable to have a choice of readings taken from any one of the six possible perpendicular directions extending outward from the center of the ball (height, width, and length measurements in either direction outward).

Now, a patented tooling ball check unit, Fig. 6, eliminates or lessens the extent of these shortcomings. Its principal feature is that the plug supporting the ball proper is removable and may be used in conjunction with any number of base pads, each of which is permanently attached to a particular tool.

Shouldered plug *A* has a 45-degree top, to which the stem of ball *B* is welded. The length of the stem is such that the center of the ball and the center line of the plug coincide. At the bottom of the plug is a shank which is a close fit in a central hole in base pad *C*.

The base pad is of two-piece construction to facilitate the milling of inspection port *D*. In addition, there is a brass retaining screw *E* in the upper piece and a height-setting screw *F* and a staking port *G* in the lower piece.

Initial setting of the height of the plug is made by adjusting screw *F*. The inspection port provides a means of checking the contact surfaces of the adjusting screw and shank, either visually or with a feeler gage.

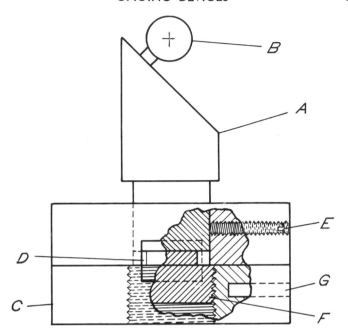

Fig. 6.   Since plug body *A* is removable, it may be used with any number of base pads *C*.

Tampering with the setting can be avoided by covering the head of the screw with solder and inserting a punch in the staking port and forcing the adjacent wall section against the thread of the screw.

The brass retaining screw secures the plug in position and cancels out the effect of any play between the plug and the base pad that might affect the accuracy of the tooling ball during a rotation of the unit. This is done by the pressure of the screw, which assures that the shank of the plug will bear against the same surface each time a reading is taken from the ball. Inspection seals applied to the heads of the two bolts used in standard practice to secure the base pad to the tool prevent all possible alterations of the original reference plane setting.

In Fig. 7 are diagrams comparing the accessible reading surfaces of a conventional tooling ball and a removable tooling ball. The arrows in views W and X indicate the accessible reading surface of a conventional tooling ball; in view W, the base pad is normal to the tool reference plane, and, in view X, it is at an angle to it. A much larger reading surface is presented by the removable tooling ball, views Y and Z. In view Y, with the base pad normal to the tool reference plane, the surface represented by the solid arrow is accessible; and by revolving the plug 180 degrees on its axis, the surface represented by the broken arrow is

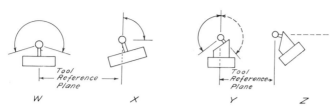

Fig. 7. In each view, the arrows indicate the extent of the accessible reading surface.

accessible. Likewise, in view Z, where the base pad is at an angle to the tool reference plane, readings can be made from the ball in any one of six directions required — with the plug in the position illustrated or revolved 180 degrees on its axis.

Comparator Gage for Internal Measurements

During investigations on creep in timber at split ring and shear plate connector joints, it became necessary to devise an instrument for recording a large number of measurements in the shortest time possible, without any great sacrifice in accuracy. The measurements were ordinarily taken with an internal screw micrometer. The limited range of this

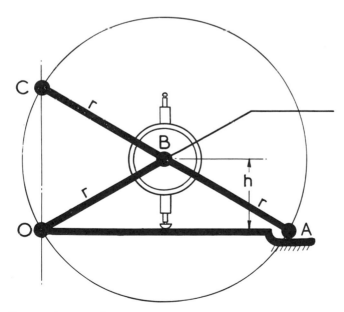

Fig. 8. Comparator gage for internal measurement is based on the Scott-Russell straight-line motion.

micrometer and the personal error in setting and reading this type of instrument made it impractical for recording as many as 640 measurements in a matter of two hours. The instrument here described was devised to eliminate the personal error in setting and to record a much greater number of readings in a limited time. It also permitted the use of a dial gage, which is much easier to read and requires less skilled operators.

The gage is of the comparator type, and is based on the Scott-Russel straight-line motion. As seen in Fig. 8, C will move on a straight line, provided A moves on a plane surface. There are two links OB and AC. OB is one-half the length of AC and is pivoted at B, which is at the mid-point of AC. OB is also pivoted at O. Since $AB = BC = BO$, a circle around B as center will intersect A, O, and C at any position of the link. Consequently, the line CO is perpendicular to OA, AOC being an angle in a semicircle.

The instrument is made of aluminum castings and, as shown in Figs. 9 and 10, has measuring points at O and C. A baseplate a, connecting O and A, contains a slot b, which provides the plane surface at A. As may be seen in Fig. 8, the travel of B is always half of that of C. An Ames dial gage No. 282, calibrated in increments of 0.001 inch, with 1 inch of travel, is suspended on a fulcrum pin and registers the movement between B and bracket c on the baseplate.

Fig. 9.　Front of gage showing slot b which provides a plane surface for sliding member A and the measuring points O and C.

Fig. 10. Back of comparator gage showing slot *d* in which extension of fulcrum pin is guided. Springs *f* keep instrument open.

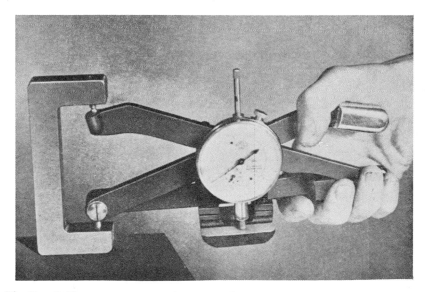

Fig. 11. Setting comparator gage by means of a standard gage piece. When dial is set at zero, readings can be made to 0.002 inch.

In order to keep the dial gage normal to the baseplate, the extension of the fulcrum pin is guided in a slot *d* which also contains a pin *e*, Fig. 10, to protect the dial gage from shock load. The springs *f* keep the instrument in the open position and facilitate its use. The initial setting of the gage by means of a standard gage piece is illustrated in Fig. 11. When the dial is adjusted to zero, the instrument is ready to be used as a comparator and direct readings can be made to 0.002 inch if the accuracy of the dial gage complies with standard specifications for this type of gage. The range of the gage can be varied by making the arms of different lengths. The instrument shown has a range of 2 to 3 1/2 inches. The travel of the dial gage obviously needs to be half of the measuring range.

Locator Compensates for Variations in Hole Center Distance

Two holes in a workpiece serve to locate it accurately over a gage during an inspection operation. While the diameters of the holes previously were machined to close tolerances, a permissible variation in the hole center distance in each piece in the lot presented a problem in positioning the work over the gage. It was solved by a tapered locator having an elliptical cross-section.

In Fig. 12, workpiece *A* resting on gage *B* has one of its holes over shouldered pin *C*. Locator *D* engages the other hole. The part of the locator entering the hole is a slight taper — 1-degree included angle — to eliminate any possible play or movement of the workpiece. Also, as can be seen by comparing the two cross-sections X-X and Y-Y, the elliptical design of the tapered part provides contact only along the short axis of the hole. Thus, the locator can enter the hole even though distance *a* varies from one piece in the lot to the next.

Finally, a radial relationship between the hole and the locator is maintained by pilot *E* which engages bushing *F* in the gage. The bottom of the pilot has a milled step which is aligned with insert *G* in the bottom of the bushing when the locator engages the hole.

Handy Bench Micrometer for Three-Wire Measurements

Three-wire measurement of screw heads is an accurate but usually cumbersome operation. If frequent inspection of thread plug gages or precision threaded work is necessary, the making of a simple bench micrometer of the type illustrated in Fig. 13 may prove to be quite a timesaver.

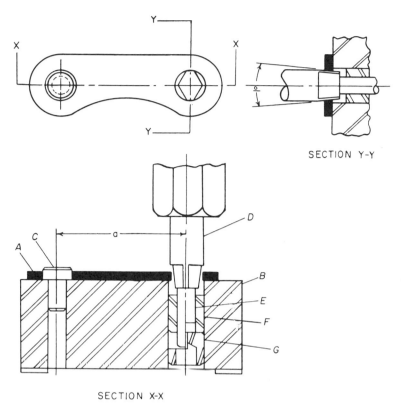

SECTION Y-Y

SECTION X-X

Fig. 12. Locator D contacts the hole along its short axis (section Y-Y), leaving clearance along its long axis (section X-X), thus allowing for variation in distance.

This measuring tool consists of three major parts, all made from bar stock. The base is carburized, hardened, and ground on all bearing surfaces, and the anvil surface is lapped to a low micro-inch finish and a high degree of flatness. A cylindrical column, which is also carburized, hardened, and ground, is pressed into a bored hole in the base. An adjustable arm equipped with a small vernier micrometer head that reads in tenths of thousandths is then fitted to the column.

To use the instrument, a stack of gage-blocks equal to the desired measurement over the work and wires is placed on the anvil. The arm is lowered and locked on the column so that the micrometer reads zero when its spindle is in contact with the stack of blocks. After the work is set up on two wires on the anvil, a micrometer reading over the third wire placed on top of the thread will show any deviation from the standard measurement as set with the gage-blocks.

Fixture for Gaging the Position of Keyways

Accurately located keyways were to be machined in a large quantity of ground shafts. In addition, the need for economy in production required that a simple and efficient method be devised for gaging both the offset and the inclination of the keyways. The device shown in Fig. 14 was used to determine these errors.

The gage has a hardened steel body A mounted on a hardened and ground steel baseplate B. Cap-screws secure the two members together and a bolt is employed at each corner of the baseplate to fasten the gage to a stand during use. A bearing hole, ground and lapped for a close fit with the workpiece C, extends through the width of the body and is parallel to the bottom of the baseplate. Across the top of the body in the center of its width is a deep rectangular slot having all sides accurately ground. This slot has its base parallel to the bottom of the gage and

Fig. 13.   Bench micrometer that is convenient for measuring screw threads by the three-wire method.

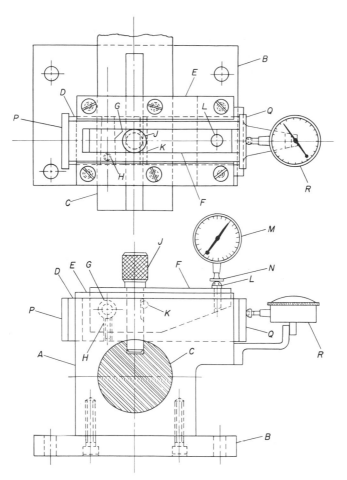

Fig. 14.   Gage used for determining the position of keyways in shafts. Both offset and inclination errors can be readily checked in a single setup.

about 3/16 inch below the top of the bearing hole. The resulting inter-section of the base of the slot and the hole is about 7/8 inch long and is centered on the vertical axis of the hole.

A steel slide D, also hardened and ground, is machined to have a T-shaped cross-section. Extending about 1/4 inch longer than the gage body, slide D is retained in the working position by two keeper plates E. Each keeper plate, being secured to the top of the body by three socket-head screws, bears lightly on the stepped sides of the slide. The slide is lapped into the horizontal slot to move freely and smoothly under light finger pressure, without vertical movement or side play.

A slot is machined through slide *D* for the major portion of its length. The width of the slot is about one-half that of the slide, but not less than one and one-half times the width of the keyway to be gaged. A rectangular arm *F* of hardened steel is lapped to fit closely in this slot. The arm is pivoted near its left-hand end on a pin *G* which is secured by a headless screw *H* in one side wall of the slide. A tapered bearing hole is reamed in the arm to suit a corresponding taper machined on the shank of the pin. The end portions of the pin which fit into the side walls of the slide *D* are cylindrical but of different diameters. The tapered surfaces of the pin and its bearing hole allow adjustment for wear, thus eliminating play between the parts. The arm, which can pivot about 2 degrees in each direction from the horizontal plane, has its top edge ground accurately flat and projects about 1/4 inch above the keeper plates.

A hardened and ground steel locator pin *J* has a shank of the same diameter as the width of the keyway in the workpiece. This pin is carefully lapped to fit smoothly in a hole in arm *F* and, at the same time, be precisely perpendicular to its top surface. Below the knurled head of the pin is a shoulder that limits the depth to which the shank enters a keyway being gaged. Member *J* is prevented from being completely withdrawn by a stop-pin K positioned across the arm slot. Pin *K* bears lightly against a flat ground for a short distance along the adjacent side of the locator pin. Secured in a hole in the right-hand upper end of the arm is a hardened steel pin *L* having a polished spherical head. The distance between the centers of pin *L* and locator *J* is about three times that between locator *J* and pivot pin *G*. Dial indicator *M* is fastened to a bracket (not shown) secured to the body and the spindle bears on the head of pin *L*. A large flat anvil *N* is fitted to the spindle to insure proper contact with pin *L* when the arm is slightly swiveled within the slide.

Rectilinear movement of the slide is limited by hardened steel plates *P* and *Q* which are fastened to opposite ends of the slide. The external side of plate *Q* is ground and lapped to provide a measuring surface for contact with the spindle of a dial indicator *R* affixed to a bracket mounted on the body.

To use the gage, a workpiece is inserted within the bearing hole in the body, and locator pin *J* is pressed down allowing the lower end of its shank to enter the keyway as seen in Fig. 14. If the keyway is vertical, but located off the vertical axis of the shaft (Fig. 15, View X), the slide can be moved horizontally in the appropriate direction to allow the locator pin to engage the keyway. Dial indicator *R* is pre-set to a shaft having an accurately located keyway; thus this indicator records any offset of the keyway.

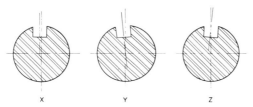

Fig. 15. Typical offset and inclination errors in the position of keyways.

In cases where the keyway is inclined in either direction relative to the vertical center axis of the shaft, as shown in Views Y and Z of Fig. 15, arm *F* will pivot within the slide and also move horizontally with it to enable the locator pin to enter the defective keyway. The amount of inclination is denoted by the reading on dial indicator *M* which is magnified because of the relative center distances of pins *G* and *L* from the vertical axis of the locator pin. Dial indicator *M* is, of course, pre-set to a master shaft having an accurately located keyway. In this way, keyway location is readily checked in one setup.

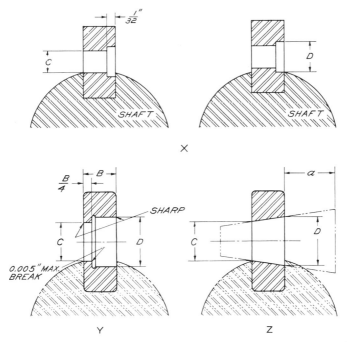

Fig. 16. Modifications of the Woodruff key-slot "Go" gage, view X, are shown in
views Y and Z.

Two Improved Designs for Woodruff Key-Slot Gages

In the accepted method of checking Woodruff key slots, the "Go" gage disc used has two bore diameters which represent maximum and minimum depth limits. As can be seen in view X of Fig. 16, the small diameter C represents maximum slot depth; and the large diameter D, minimum depth. Gage-disc dimensions are given in MACHINERY'S HANDBOOK. A tolerance of plus 0.0004, minus 0.0000 inch is required for each bore diameter, with the length of diameter D fixed at 1/32 inch for all gage-disc sizes.

The shallowness of diameter D makes it difficult to be inspected with ordinary measuring instruments or plug gages, when manufacturing the gage. For this reason, two alternate gage-disc designs, shown in views Y and Z, are proposed.

In view Y, the length of diameter C is shortened to one-fourth of the disc width, but not to less than 0.020 inch. The greater length of diameter D can thus be readily checked with a plug gage or a toolmaker's microscope. (When this diameter is only 1/32 inch long, light would be diffused when striking a radius in the hole.)

The second design, view Z, consists of taperboring the disc. The degree of taper is determined by the difference in the diameter of the bore at the two faces of the disc. A plug gage is first made to this taper, and by measuring the distance a, it is possible to control the size of the bore.

Gages for Checking Depth of Keyways

Two easily made gages, one for checking the bore and keyway in a part and another for checking the depth of keyway in a shaft, are shown in Fig. 17. These gages provide a simple means of checking the parts to assure cutting the keyways to the correct depth and finishing the bore to fit a shaft without looseness when the parts are assembled.

The gage for checking the diameter of bore and depth of keyway consists of three parts — the tool-steel plug C, a knurled handle A, and the keyway depth gage B. Plug C is drilled and tapped to receive handle A, and has a slot machined in it at an angle of about 3 degrees with its axis. This slot is finished to a nice sliding fit for keyway depth gage B.

Plug C is hardened and ground on the surfaces indicated. The cylindrical portion is ground to a close sliding fit for a bore of the correct diameter to give the desired fit when assembled on a shaft of specified size. The keyway gage is also made of tool steel, and is hardened and ground on the surfaces indicated.

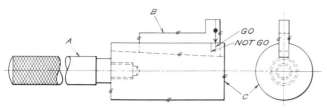

Gage for checking diameter of bore and depth of keyway

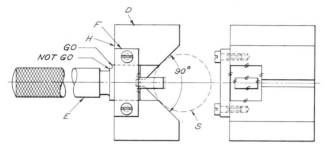

V-Block and flush pin gage for checking depth of keyway
in shaft shown by dotted lines at S

Fig. 17.   Gages designed to facilitate machining bores and keyways to insure close fits
when keys are fitted in shafts assembled in bores.

In using this tool, the plug *C* is first inserted in the bore to be gaged. To pass inspection, the bore must be just large enough to admit the plug with a close gaging fit. Assuming that the bore passes inspection, the depth of the keyway in the bore can next be checked by sliding gage *B* into place until it is a close wedge fit between the bottom of the slot in plug *C* and the bottom of the keyway in the bore. If the arrow point scribed on the side of gage *B* coincides with either the "Go" or "Not Go" line scribed on the cylindrical surface of plug *C* or is located between these lines, the depth of the keyway is within the required tolerance, and the part is passed as satisfactory.

The gage shown in the lower view of Fig. 17 consists of a V-block *D*, a flush pin type keyway depth gage *E*, and a plate *F* secured to the V-block by cap-screws. The V-block *D* serves to locate the gage in the correct position on the shaft for gaging the depth of the keyway. Flush pin *E* has a knurled handle, and is hardened and ground. It has a rectangular section with "Go" and "Not Go" steps and a tongue at the end which is accurately ground to a sliding fit for the keyway to be gaged.

The rectangular section of the flush pin is ground on all four sides to a close sliding fit in a slot in the end of the V-block. The gaging tongue and rectangular section of the flush-pin are accurately centered, so that the gaging tongue will slide into an accurately machined keyway in the shaft when the V-block is in place on the shaft. Plate *F* serves to hold the flush pin in the slot in the V-block.

In use, the gage is placed on the shaft indicated by dotted lines at *S*. The flush pin *E* is then pushed in until the end of the gaging tongue is in contact with the bottom of the keyway in the shaft. In order to pass inspection, either the "Go" or the "Not Go" step on the flush pin must be flush with surface *H* of the V-block, or surface *H* must be located between the "Go" and "Not Go" steps.

Simple Gage for Length Dimension

Recently it was necessary to design a gage for checking distance *X* (see Fig. 18) from one side of a milled groove in a part to an adjacent end of the work. The dimension had to be held within ±0.005 inch.

A simple gage was designed with a base *A* that has a flat top surface on which the workpiece can be firmly placed. A hardened and ground insert *B* provides a means of registering the critical side of the work groove. The distance from this point of registry to a shoulder on the right-hand end of the gage base is equal to the lower limit of dimension *X*.

Knurled collar *C* has a slip fit on a shank at the end of base *A* and also a slip fit between the shoulder of the base and collar *D*. Step *E*, 0.010 inch wide, was ground across the upper portion of the collar on the side adjacent to the base, this dimension being equal to the permissible tolerance of dimension *X*.

In an inspection, if it is possible to lay the workpiece on the gage as shown, it is indicative that dimension *X* is within the maximum limit. Then knob *C* is swiveled, and if it is impossible to turn the knob around completely because of step *E* contacting the end of the work, it is indicated that the work is longer than the minimum limit of dimension *X*. Parts that are too long cannot be loaded on the gage because of inter-

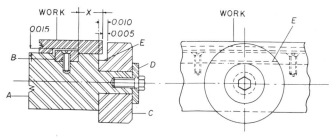

Fig. 18. Simple "Go" and "Not Go" gage for accurately checking the distance from a groove to the end of a workpiece.

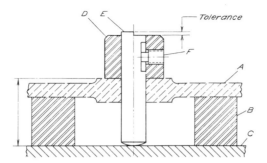

Fig. 19.   The height of the boss is within tolerance if top of body *D* is between, or
flush with, the step or end of pin *E*.

ference with the face of collar *C*. Conversely, parts that are too short
are indicated when the knob can be rotated without interference.

Boss Height Inspected with Flush-Pin Gage

A boss on a gear box top must be machined to a specified height to
provide clearance for a gear-shaft assembly. To inspect this dimension,
a flush-pin gage is used. In Fig. 19, the work *A* is supported by a ring *B*
over a surface plate *C*. On top of the boss is the gage body *D* containing
the pin *E*. One-half of the upper end of the pin is ground to form a step.
The distance from the end to the step is made equal to the boss height
tolerance.

Over-all length of the pin is such that when the work is on ring *B*, the
step will be flush with the top of the gage body at the lower limit of the
tolerance on boss height; and the end of the pin will be flush with the
top of the gage body at the upper limit of the tolerance on boss height.
A set-screw *F* in the side of the gage body is contained in a slot in the
pin, and keeps both members together when the gage is handled.

Gage for Measuring Conical Recess

A special telescoping gage designed to facilitate checking the mean
diameter of a conical recess machined in a cast-iron part is shown in
Fig. 20. Although specifically designed for use in measuring the diameter *Y* of the conical recess in the piece *W*, the gage can also be adapted
for other checking operations, such as measuring dovetail grooves. The
difficulty or impossibility of accurately measuring the diameter of the
conical recess with instruments ordinarily available led to the development of the telescoping gage. Internal micrometers and gage-blocks, for

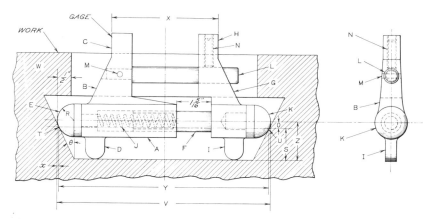

Fig. 20.   Gage of special design developed to permit accurate measuring of conical-
shaped recess by means of an outside micrometer.

example, could not be inserted or properly located through the bored
hole above the conical recess, a difficulty which was overcome by the
telescoping feature of the special gage.

In Fig. 20 the telescoping gage is shown in the correct position for
measuring the mean diameter Y of the conical recess in piece W. When
in this position, the half-spherical buttons E and K make contact on
diametrically opposite sides of the conical recess, or cavity, at points
T and U. The lugs D and I are finished to the accurately calculated
length Z, which equals $S + O$, where S is one-half the height of the
conical recess and O equals the offset required to bring buttons E and K
into contact with diametrically opposite sides of the conical recess on
the mean diameter line. The amount of offset O (see Fig. 21) equals the
radius R of the contact button multiplied by the sine of the angle of
inclination θ of the side of the recess.

The arms B and G, Fig. 20, are so machined that the dimension X,
measured over the faces C and H, is exactly one-half the dimension V,
measured over the ends of contact buttons E and K, when the gage is
collapsed or adjusted so that V is equal to the dimension Y to be gaged.
In this example, the actual mean diameter Y to be checked is 8 inches.
Therefore, when the gage is set so that V is 8 inches, the corresponding
measurement X over faces C and H is 4 inches.

However, when the gage is used to check dimension Y (which is 8
inches) the actual micrometer reading will be 4 inches plus $2x$. When
thus set, the gage has a collapsing range of 1 5/16 inches, which allows
a clearance of 5/32 inch between each button and the bore of the large
hole above the conical recess. This is sufficient to permit insertion or

removal of the gage when it is fully collapsed. Making the gage with dimension X equal to one-half Y, as described, facilitates accurate checking of the gage itself and simplifies the calculation of measurements required in designing, constructing, and using the gage.

The diagram Fig. 21 shows an enlarged cross-section of the half-spherical contact button K in the measuring or checking position — in contact with the side of the recess in the work. Referring to this diagram, the calculations required in designing and using the telescoping gage can be made by using the following formulas:

$$O = R \times \text{sine } \theta,$$

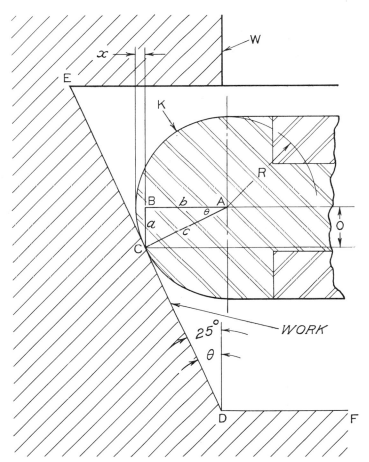

Fig. 21. Diagram illustrating method of making calculations required in designing, constructing, and using gage shown in Fig. 20.

in which

O = amount center of contact buttons E and K are offset above mean
        diameter line to be measured in order to have buttons make
        contact with sides of conical recess on the mean diameter line;

R = radius of the spherical contact buttons E and K;

θ = angle of inclination of the side of conical recess.

The amount x, which must be doubled and added to 4 to obtain the
correct micrometer measurement X, Fig. 20, in inches, for a correct
measurement of 8 inches for Y, is found by the formula

$$x = R - (R \times \cos \theta)$$

The telescoping feature, which permits the gage to be collapsed for
insertion in the cavity, then expanded for measuring, and finally collapsed for removal, is obtained by having the shank F, Fig. 20, a close
sliding fit in a lapped hole in the body A. The guide rod L, secured in
arm B, is a close sliding fit in a lapped hole in sliding arm G, and serves
to keep the sliding member with locating button K in accurate axial
alignment with button E and measuring face H in accurate alignment
with measuring face C. A brass retaining screw N is provided to permit
locking the sliding arm in any position desired.

In applying the gage, screw N is first loosened and shank F of sliding
arm G pressed into the bore of body A and arm G brought in contact
with arm B, compressing the light spring J. Screw N is then tightened
to hold the gage members in the collapsed position. When thus collapsed, the gage is lowered into the recess until lugs D and I rest on the
bottom of the conical recess. With the gage in this position retaining
screw N is loosened. Spring J will cause the sliding shank F to move
outward until button K makes contact with the side of the recess. By
carefully moving one end of the gage slightly to and fro in a lateral
radial direction, the gage can be easily set to the maximum measuring
diameter on the mean diameter of the conical recess. In this position,
the gage should be set by the retaining screw N. Having thus set the
gage, a micrometer reading is taken over faces C and H to obtain measurement X for comparison with a predetermined measurement for this
dimension, which is either calculated or obtained by the cut-and-try
method from a sample component to correspond with the correct mean
diameter Y of the recess. After the measurement X is taken, the retaining
screw N is loosened, the gage collapsed, and the screw N tightened.
The gage can then be removed, ready to measure another piece.

All the critical working elements and measuring contact points or
buttons of the gage must, of course, be hardened and accurately fitted.

The gage can be used to measure annular grooves and slots of various kinds which are in accessible positions.

Comparator for Checking Tapers

Production checking of tapers can be speeded with the comparator illustrated in Fig. 22. The device, which measures the change in diameter of the taper in a given axial distance, is easily made and quick and simple

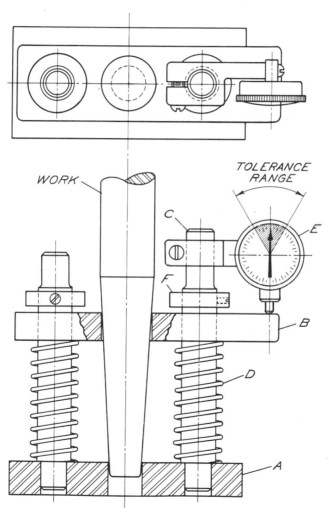

Fig. 22. The taper is acceptable if the indicator reads in the tolerance range.

in operation. It is superior to a sleeve gage, in that no bluing is required.

The comparator consists of a baseplate *A*, bridge-piece *B*, a pair of posts *C* and springs *D*, and a dial indicator *E*. The posts, joining the baseplate and the bridge-piece, are a press fit with the baseplate and a sliding fit with the bridge-piece. The springs enclose the posts, keeping the bridge-piece in normal contact with stop collars *F*.

One end of the bridge-piece is somewhat longer, forming an anvil for the indicator, which is strapped to the top of one of the posts. Both the baseplate and the bridge-piece are bored cylindrically, each to a size corresponding to the known diameter of the required taper at a point along its axis.

In operation, a tapered master is inserted in the bridge-piece, which is depressed until the end of the master makes contact with the baseplate. With the bridge-piece still depressed, the indicator is adjusted to read zero. Workpieces are then substituted for the master, to which they are compared by observing the indicator. Acceptable tapers will return the indicator to a tolerance range that can be established on either side of the zero reading.

Useful Shop Gage for Accurate Checking of Tapers

The gage shown in Fig. 23, together with an ordinary outside micrometer, is all that is needed to quickly and accurately determine the taper

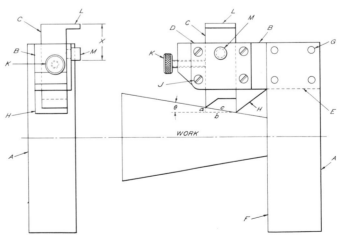

Fig. 23.   Simple gage for use with a micrometer to measure the taper per inch of a surface.

per inch of a frustum of a cone, such as a tapered plug gage. The only requirement is that at least one of the bases of the component being measured be square with its center line.

The principal parts of the gage are a beam A, an arm B, a movable finger C, and a keeper-plate D. All these parts are made of steel. The beam can be of any suitable length. It is a thicker member than the arm, which it accommodates in a milled slot E. This slot is made exactly 90 degrees with the register edge F of the beam. Four rivets G retain the arm permanently in the slot.

On the under side of the arm and immediately adjacent to the register edge of the beam is an integral extension H exactly 1.000 inch wide. Its lower end is ground back to create a sharp edge with its left side, and immediately adjacent is the movable finger C, also exactly 1.000 inch wide. This finger is free to slide up or down in a guideway in the arm against the friction of the keeper-plate, held in place by four screws J. By tightening a thumb-screw K in the end of the arm, the setting of the finger can be fixed. Like the integral extension, the finger is ground back at its lower end at an angle to create a sharp edge with its left side. At its top, the finger has a lip L projecting over the keeper-plate.

A short cylindrical dowel pin M is pressed into the face of the keeper-plate, with the distance X from the bottom of the pin to the top of the lip measuring exactly 1.000 inch when the lower edge of the finger is in the same horizontal plane as the lower edge of the extension. (In constructing the gage, this distance can be established by first leaving the lip thicker than is required, then aligning the two edges with a try-square and grinding off the top of the lip until the distance X is exactly 1.000 inch. Or, the diameter of the pin can be left over size, then removed and ground down the necessary amount.)

In order to determine the taper per inch, the finger is raised and the work held in the position illustrated — the extension in contact with the tapered surface, and one base of the work bearing against the register edge of the beam. As has been stated, this base must be known to be square with the center line of the work. The finger is next lowered into contact with the tapered surface, and fixed by tightening the thumb-screw. The work can then be removed.

Distance X is read with a 2-inch micrometer. By subtracting 1 inch from the reading, the height of the side a of the right triangle abc is derived. This distance is also the taper per inch of the surface.

It is then a simple matter to establish the angle θ which the tapered surface forms with the center line of the work. For example, if the micrometer reading for distance X is 1.176 inches, then a = 0.176 inch, and

$$\text{Tan } \theta = \frac{a}{b} = \frac{0.176}{1}$$

By referring to a table of natural trigonometric functions, it is found that $\theta = 10$ degrees. The included angle of the component under consideration is 20 degrees, and the included taper per foot is $0.176 \times 2 \times 12$, or 4.224 inches.

If the component had been gaged in a reverse manner, with its larger base adjacent to the register edge, then the distance X would be measured with a 1-inch micrometer, and the height of the side a would be obtained by subtracting the micrometer reading from 1.000 inch.

Lead-Screw Adapted for Gaging Long Shafts

A handy arrangement for accurately locating shoulders and grooves in long shafts being turned in a lathe is illustrated in Fig. 24. The device, which consists simply of an attachment for counting the number of complete and partial turns made by the lathe lead-screw, is used to gage the carriage travel. This method eliminates the need for the long, specially made gaging rods.

A dial A is attached by a thumbscrew to the tailstock end of the lathe lead-screw B. This dial is calibrated in length of carriage travel, the number of divisions being equal to the pitch of the lead-screw in thousandths of an inch. The dial shown has 250 divisions. If its calibration is satis-

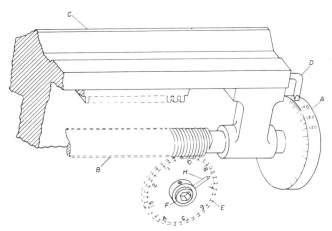

Fig. 24.   Dials operated by lathe lead-screw that accurately measure carriage travel and are handy for turning long shafts.

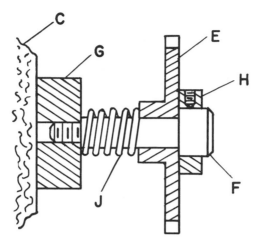

Fig. 25. Method used for attaching dial *E* to the lathe bed *C* shown in detail. Spring *J* allows easy resetting of the dial.

factory, a milling machine dial can be used for this purpose. The end of the lathe bed *C* is drilled for a pointer *D* to permit easy reading of the dial. Resetting this dial to zero is a simple matter since it is secured with a thumbscrew.

Although dial *A* and a scale are all that are necessary to measure carriage travel to the 0.001 inch, a second dial *E*, calibrated in inches, may be added for convenience. Actually, dial *E* is a narrow-face gear of a suitable pitch to mesh with the lead-screw. This gear is mounted on a shoulder screw *F* secured in a block *G* (Fig. 25), which, in turn, is attached to the lathe bed. Screw *F* is oriented at a slight angle to the axis of the lead-screw to permit dial *E* to mate properly with the lead-screw. In addition, a pointer *H* is secured to screw *F* by a set-screw. To reset dial *E*, it is pushed in, depressing a spring *J* (Fig. 25) mounted behind the dial on screw *F*, and rotated to zero. Some lead-screws have keyways and are used for feeding as well as threading. In such cases, the portion of the keyslot that would interfere with dial *E* may be filled with solder and the solder filed in the shape of the threads. On most lathes, the key never reaches this end-portion of the lead-screw.

When turning long shafts, the regular feed-screw and a scale can be used to measure the roughing cut. Then, by changing over to the lead-screw for the carriage drive, the length of the finishing cut can be measured with dials *A* and *E*. A feed sufficiently fine for most finishing can be obtained by shifting the change-gears to cut the maximum number of threads per inch available on the lathe. If this is not satisfactory, dials

A and *E* can be used in cutting the shoulders to length and the peripheral surfaces can then be finished with the usual fine feed. This attachment should not be used when the feed-screw is employed since the clutch may slip, destroying the accuracy of measurement. When the dials are in the position shown, the carriage has traveled 7.397 inches.

Height Comparator Employs Standard Micrometer Head with Special Body

A quick-setting height comparator of versatile use in the machine shop, tool-room, and inspection department is seen in Fig. 26. It incorporates a commercially available micrometer head with a specially designed body having a series of measuring surfaces spaced 1.000 inch apart.

Fig. 26.   The height being measured can be transferred to the comparator by means of a dial indicator.

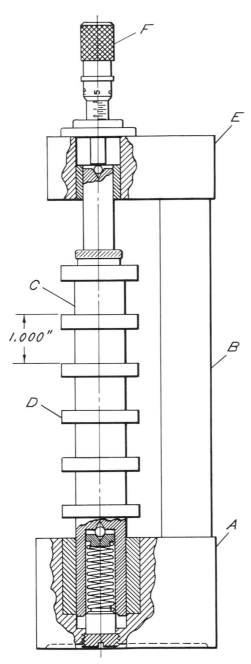

Fig. 27. The series of measuring surfaces *D* enables a single micrometer head *F* to measure any height within the range of the comparator.

In Fig. 26 the man at the right is showing how the device can be used to measure the height of a surface. First, a surface gage on which is clamped a dial indicator is positioned on the reference base of the object and the indicator is set to read zero at the height to be measured. Next, the surface gage is positioned near the comparator, which is then adjusted by turning the thimble of the micrometer until the indicator again reads zero over one of the measuring surfaces.

With the micrometer fully closed, each of the measuring surfaces is at a precise inch-unit distance above the base of the comparator. Thus, by adding the reading of the micrometer head to the inch distance of the measuring surface on which the indicator bears, the height is obtained.

Alternately, the comparator can be set at some desired height, and the indicator used to transfer the reading to the work or tool. A vernier on the micrometer head permits settings to be made to 0.0001 inch.

The design of the comparator can be seen in Fig. 27. It consists of base A, a support column B, a post C, and a crown E which contains the

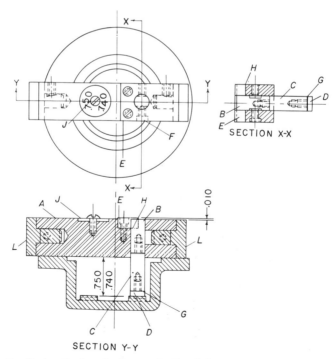

Fig. 28.   By altering the length of gage pin C and the thickness of tolerance spacer H, this depth gage can be used in many different applications.

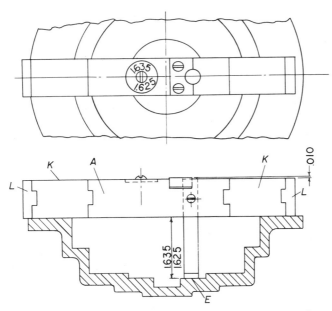

Fig. 29.   Further alteration of the basic gage bar *A* can be made by adding one or more extension units such as those indicated at *K*.

micrometer head *F*. The spindle can move vertically in the crown and base. In the bottom end of the post, which is hollow, is a spring to maintain the post in contact with the end of the micrometer spindle.

Building-Block Depth Gage

The depth gage shown in Figs. 28 and 29 is both unique in design and universal in application. By adding or removing sections, the gage body can be lengthened or shortened in a building-block manner to permit use with a wide variety of workpieces. Many different depths can be checked with this tool by replacing the gage pin with one of appropriate length.

Distance between the face of an internal boss and a mounting flange is shown being checked in Fig. 28. Gage bar *A* is drilled to receive the gage-rod assembly consisting of a gage-pin adapter *B*, gage pin *C*, and contact tip *D*. The gage pin has a threaded shoulder on one end and a tapped hole in the other end so that the guide-rod assembly can be altered quickly.

A tolerance block *E* is seated in a recess in the top face of the gage bar. When tightened down by two screws, the upper face of the tolerance block is flush with that of the gage bar.

The depth dimension to be checked in the illustrated case is 0.750 plus 0.000 minus 0.010 inch. To set the gage for this reading, the top end of the gage-pin adapter is placed flush with the upper face of the gage bar and locked with set-screw *F*. A proper-sized gage pin and a contact tip are then assembled to the adapter. If the distance from the lower surface of the gage bar to the lower surface of the contact tip is not quite 0.750 inch, a spacer *G* of adequate thickness can be used.

To allow for the 0.010-inch tolerance, a spacer *H* of this thickness is placed beneath tolerance block *E*. The gage can be identified with the setting of the gage rod by stamping the tolerance range on a thin replaceable nameplate *J*. Upon release of set-screw *F* the gage is ready for use.

As shown in Fig. 28, the gage bar is positioned across the flange of the workpiece. The top of the gage rod is then pressed down firmly so that contact tip *D* rests solidly on the boss to be checked. If the work has been machined correctly, the upper end of the gage-pin adapter will be located between the upper face of the tolerance block and the upper

Fig. 30. Chips and cutting oil are cleared from holes during checking by means of air jets provided in special plug gage. Gage, shutoff valve, and hose are connected to shop air line.

face of the gage bar. An offset tip can be easily made up to check other-wise-inaccessible internal surfaces.

An example of the building-block nature of this gage can be seen in Fig. 29. Here, a short extension K has been added to each end of the gage bar A to accommodate a large-diameter workpiece.

To permit the joining of additional units, the ends of the gage bar are blind-drilled and have shallow slots milled across them. This hole-and-slot arrangement receives protector caps L (Figs. 28 and 29) which are always placed at the ends of the gage. The caps are secured by set-screws.

Each extension bar has a drilled hole and a milled slot at one end, and a turned protrusion and a milled tongue at the other end. In this way, any number or size of extensions can be added to the main gage bar.

Air Jet in Plug Gage Clears Holes

The production job shown in Fig. 30 involves the drilling and reaming of a large number of close-tolerance holes through a group of laminated plates. To adequately control tolerances, 100 per cent inspection was necessary as machining progressed. Checking of the holes was greatly facilitated by special "Go" and "Not Go" plug gages which were pro-vided with air-jet holes in the end faces and also radially in the side walls. The gages, plus shutoff valves, were attached to hoses connected to the shop air line (see Fig. 30). Air pressure of about 50 psi was found sufficient to effectively clear the holes of all chips and cutting oil when the gage was inserted.

CHAPTER 12

Tooling for Layout

Enlarging Tool Speeds Layout of Master Cams

Designers of automatic machinery will appreciate this time-saving layout tool which eliminates hours of work when it becomes necessary to enlarge cams. It is commonly the case that an oversize master cam, to be used on a cam-cutting machine, must be made up from a full-scale cam drawing. With the tool shown in Fig. 1, the enlargement can be laid out directly from the original drawing without depending on radial lines or resorting to the use of a compass.

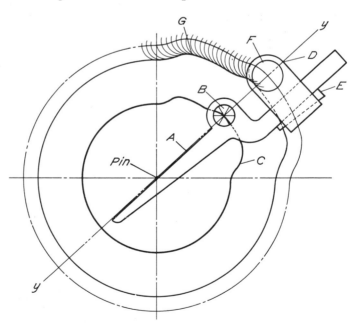

Fig. 1. Layout tool permits rapid and accurate enlarging of cam drawings. It is especially useful when large master cams must be made to guide cam-cutting machines.

The body of the layout tool should be made of either aluminum or stainless steel. Edge A of the lower end of the tool is absolutely straight and rests on line Y-Y, which passes through the center of the cam to be reproduced. This edge is maintained in alignment by holding it in contact with a pin placed at the center point of the cam drawing.

A clear plastic marker B is inserted in a hole at the upper end of edge A. Six equally spaced radial lines are provided across the face of the marker — their point of intersection falling on line Y-Y. It is this point that is used to sight and follow contour line C, representing the path to be traced by the center line of the follower-roller.

Adjustable head D slides on the offset upper end of the layout tool. The head can be locked to the tool at any point within its range of movement by means of wedge E. Positioning of the head determines the degree of cam enlargement that will be obtained.

A 2-inch hole F in the adjustable head serves as a guide for tracing the outline of the roller to be used on the cam-cutting machine. The center of this hole is located exactly on line Y-Y, thus assuring an accurate mathematical relationship between locating edge A, marker B, and the roller outline traced through hole F.

Reducing inserts can be made up to permit the tracing of circles representing smaller-size rollers. The inserts should have a 2-inch outside diameter to fit in the original hole, and an inside diameter machined to suit the roller on the cam-cutting machine. A suggested assortment of reducing inserts would include the following sizes: 1 3/4, 1 1/2, 1 1/4, and 1 inch. It is advisable to have the inside diameter approximately 1/32 inch oversize to compensate for the pencil line. The enlarged cam outline G is plotted in the usual way, that is, by a line connecting the series of traced circles.

Indexing Fixture for Locating Bolt Holes

High-vacuum systems designed for research application require the use of many bolted flanges to facilitate assembly, cleaning, and inspection. To simplify laying out the bolt-hole circles and the individual hole center lines, the fixture shown in Fig. 2 was constructed. An 0.500-inch diameter hole is first drilled and reamed through the center of the flange to be laid out. The flange is then placed over stud A of the fixture. A micrometer, set to the radius of the bolt-hole circle plus an additional 0.500 inch, is used to position plunger B. The 0.500 inch is added to the radius dimension because both the stud and the plunger are 0.500 inch in diameter.

Fig. 2. Fixture permits indexing and center-punching of equally spaced bolt holes located on a common circle.

After the plunger has been accurately positioned, it is locked in place by tightening knob C at the split end of support arm D. Plunger B is now in position to be used as a center-punch.

Washer assembly E is placed over stud A and screwed in place. This clamps the flange blank to index-plate F which is secured to a 180-tooth, 32-pitch gear G. The gear teeth are numbered and, by means of a single index-finger mounted in housing H, the following groups of equally spaced divisions can be obtained: 2, 3, 4, 5, 6, 9, 10, 12, 15, 18, 20, 30, 36, 45, 60, 90, and 180. With the addition of a second index-finger accurately located to position gear G at one-half tooth intervals, this range would be increased.

Layout Tool Scribes Large Arcs Accurately

The establishment of a series of calculated points, or points developed by geometrical construction, are the usual methods of laying out large arcs whose centers are outside the work. Either method is time-consuming and the resulting curves are not always true arcs. The same is true when it is inconvenient or impractical to establish auxiliary centers.

Figure 3 illustrates a layout tool for scribing this type of large arcs. The device consists of two L-shaped members pivoted at point a, and provided with a clamp to lock them at the necessary angle θ, for scribing the arc. A scriber is located at the pivot point, which is in line with an edge of the long legs. The short legs have buttons, located at right angles to the long leg, used for setting the legs at the required angle Θ. Pivot-pins, having ground notches extending exactly to their centers, act as bearings for the long legs when scribing the arc. They are 0.500 inch in diameter, have a slip fit in the bushings shown, and each is grooved at its lower end for a retaining pin. As the radius of the bushing body plus the radius of the legs at the pivot point is approximately 1 inch, in the case illustrated, the bushings are positioned on the extension of the arc at a distance greater than this from the points *b* and *c*, or at points *d* and *e*, on chord *C*. This is necessary in order to insure that the scriber pass somewhat past the points *b* and *c*. A shoulder is provided on the bushings for clamps (not shown) used in securing them to the work. One wall is partly cut away (final operation) for clearance of the long legs. Pins in their lower ends retain the pivot-pins.

The two identical L-shaped members are band-sawed from 1/4-inch flat-ground stock, 6 inches wide and 12 inches long. To assure pair identity, the two blanks are matched and soldered together on their flat faces. Handled as one piece, their outside contours are smoothed and finished. The holes are drilled. The solder is next melted out to separate the pair, which are then separately heat-treated and straightened. Both flat faces are "cleaned-up" by light cuts of a surface grinder.

Finishing operations call for a second matching of the L-shaped members and soldering, after which they are finish-ground on the

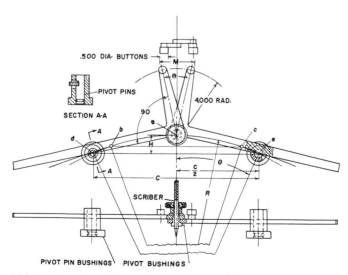

Fig. 3. Using this novel scribing tool, accurate arcs can be laid out such as that bounded by radius *R* between points *b* and *c* with a given chord. Critical dimension *H* is developed from a standard formula.

contoured edges. The holes are jig-ground. Then they can be finally separated.

Both L-shaped members are identical — they become right- and left-hand at assembly. Their dimensions are, of course, arbitrary. In the tool shown, the long legs are 12 inches and have the 0.500-inch diameter buttons located in the short legs 4.000 inch from the pivot point. The center bushing, a press fit in the lower leg and a slip fit in the upper leg, has its end threaded for the clamping knob. The center hole is a friction-tight fit for the scriber.

The soldered pair method of construction allows important dimension to be held more easily. Time is saved in not having to handle two different pieces. The leg length, of course, is arbitrary. The 12-inch length is convenient. Occasionally an arc with a longer chord has been needed. It is scribed by establishing additional positions for one of the bushings further out on the legs. It is only necessary to solve two right triangles to establish these locations. One bushing is left in its original position while the other is placed at the new position. The first one is then moved to its new location. The outside diameter of the 0.500-inch diameter pivot-pins is used for locating the bushings.

The following calculations, necessary for locating the bushings and setting the leg angles, are easily made. An example is illustrated in the sketch where a 25.000-inch radius R is required to be scribed tangent to two lines at points b and c, and at a chordal distance of 7.500 inch.

For locating the pivot-pin bushings at points d and e, an easy-to-work-with figure of 10.000 inch for chord C is assumed. The angle θ for setting the arms is:

$$\sin \theta = \frac{C}{2R} = \frac{10.000}{50.000} = 0.2000; \text{ and } \theta = 11°32.2'$$

The measurement across the buttons is found from:

$$M = 0.500 + 8 \sin \frac{\theta}{2} = 0.500 + 8 \times 0.10051 = 1.3043$$

The dimension H for the other location of points d and e can be found from the standard geometrical formula:

$$H = R - 1/2\sqrt{4R^2 - C^2}$$

but a much easier way to determine the actual dimension H is to use versed sines that are given in MACHINERY'S HANDBOOK:

$$H = R \text{ vers } \theta = 25.000 \times 0.02020 = 0.505$$

With the necessary dimensions for setup established, the bushings are clamped in position, and the legs kept in contact with the notched pins while the arc is scribed.

V-Block with Tilted-V Design

A toolmaker's V-block having several unique features that facilitate laying out precise workpieces is illustrated in Fig. 4. When a part is held on this inspection and layout aid, the exact position of the workpiece in relation to either of the two reference sides of the V-block is always known.

For convenience, the V-groove is oriented at a 45-degree angle to that of a standard V-block. With this arrangement, one side of the V-groove is vertical, the other, horizontal. As this V-block is a precision tool, all surfaces are accurately ground flat and parallel, or square, to each other.

The second unusual and important design feature of the V-block is that the length of its various sides are in exact multiples of an inch. Dimensions of the V-block shown in the illustration are 2 by 3 by 4 inches, the sides of the V-groove being 1 inch wide. This permits the V-block to be used as a size block for a 1-, 2-, 3-, or 4-inch dimension. These sizes, however, may be varied to suit the work at hand. Importance

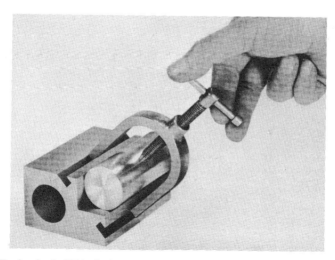

Fig. 4. Toolmaker's V-block that has V-groove oriented to facilitate laying out and inspecting workpieces. Each side of the V-groove is parallel to one of two possible reference (or base) surfaces.

of the design for larger size V-blocks should not be overlooked. A special U-shaped clamp which mounts in the two slots parallel to the V-groove is used to secure a workpiece to the V-block. This design permits either of two sides of the V-block to be used as a base without resetting the clamp. In addition, each end may also be used as a base.

To use the V-block, the workpiece is secured in the manner illustrated. Since the two base sides are parallel to the sides of the V-groove, layout and inspection of surfaces at right angles are accomplished simply by flipping the V-block from one base to the other. Also, the height for setting layout and measuring tools is obtained directly by adding the known distance between the base and the side of the V-groove to the required dimension. The accurate heights of the various sides of the V-block are handy for size reference and for supporting work in a precise position. This V-block is especially advantageous when used in combination with a sine plate, both as a size block and as a work support.

A Common Problem in Gage and Die Design

The construction of a circle that is tangent to the base of a triangle and also to the other two sides extended beyond the base is a frequent requirement in die and gage design.

In Fig. 5, triangle ABC is shown with sides CB and CA extended and the exterior angles at B and A bisected. The intersection of these angle bisectors at point O is the center of the required tangent circle. The problem of determining OT, the radius of the tangent circle, may also be solved by trigonometry.

In triangle ABC of Fig. 6, side AB with angles a and b are given. To find R, the following step-by-step analysis may be used:

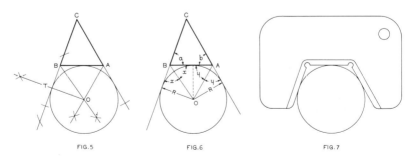

FIG.5   FIG.6   FIG.7

Fig. 5. Geometric construction used to determine the center of a circle tangent to a side of a triangle and also to the other two sides extended; Fig. 6. Diagram used to find the radius of the tangent circle by trigonometry; and Fig. 7. Checking the correctness of this gage is facilitated by the use of Formula 5.

(1)
$$\text{Angle } x = \frac{180 - a}{2} = 90 - \frac{a}{2}$$

(2)
$$\text{Angle } y = \frac{180 - b}{2} = 90 - \frac{b}{2}$$

(3)
$$OB = \frac{R}{\sin x}; \text{ and } OA = \frac{R}{\sin y}$$

(4)
$$AB = OB \cos x + OA \cos y$$

$$= \left(\frac{R}{\sin x}\right) \cos x + \left(\frac{R}{\sin y}\right) \cos y$$

$$= R \cot x + R \cot y$$

$$= R \cot \left(90 - \frac{a}{2}\right) + R \cot \left(90 - \frac{b}{2}\right)$$

$$= R \tan \frac{a}{2} + R \tan \frac{b}{2}$$

$$= R \left(\tan \frac{a}{2} + \tan \frac{b}{2}\right)$$

or,

(5)
$$R = \frac{AB}{\tan \frac{a}{2} + \tan \frac{b}{2}}$$

The use of Equation 5 is demonstrated in the following example:

Let $AB = 1$ inch, $a = 70°$, and $b = 60°$.

$$\text{Tan } \frac{a}{2} = \tan \frac{70°}{2} = \tan 35° = 0.70021$$

$$\text{Tan } \frac{b}{2} = \tan \frac{60°}{2} = \tan 30° = 0.57735$$

$$R = \frac{1}{0.70021 + 0.57735} = 0.7827 \text{ inch.}$$

A gage of the type shown in Fig. 7 is a typical case in which Formula 5 can be applied. The correct angular relationship of three working surfaces is established by means of a sine bar. A disc (the diameter of which has been determined by this formula) is used with the gage to determine any inaccuracies of the linear dimensions. Discrepancies may be checked by using a comparator.

CHAPTER 13

Die and Component Designs

Inverted Punches for Progressive Bending Dies

Progressive dies are often designed to incorporate one or more stations for forming or bending. These operations are commonly performed by an upper punch A, which bends a portion of the workpiece B downward into a recess in a stationary die-plate C, as seen at X in Fig. 1. If bending is accomplished in this manner, automatic feeding is sometimes made difficult or impossible because of the height to which the strip must be raised for each advance.

This problem can be eliminated by arranging the die to bend the work upward as shown in view Y. A lower punch D, located in the die-holder, is actuated through a lever E by a member attached to the ram of the press. Lever E simply converts the downward movement of the ram to an upward movement of the punch. The bending is done against a punch F attached to the upper part of the die. A suitable compression spring (not shown) effects the return of the lower punch.

The same setup may be used to form a U-bend after a simple right-angle bend has been made, at a previous station. However, a somewhat more complicated upper punch G (view Z) must be employed. This punch has an additional pivoting member H which ordinarily hangs in

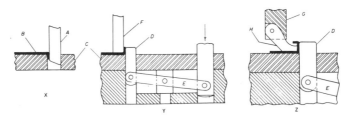

Fig. 1. Automatic feeding of strip stock in progressive dies can be facilitated by forming bends upward.

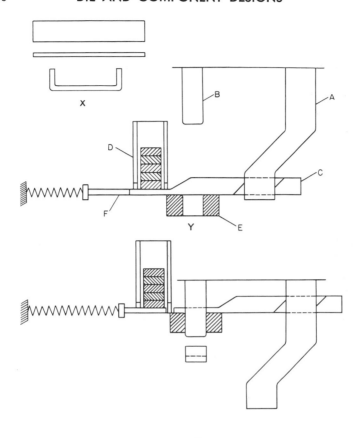

Fig. 2.   On the down stroke, cam *A* pulls slide *C* into the press. Then punch *B* pushes the blank through die *E*.

a vertical position. When the press ram descends, this member is rotated to a position against the workpiece, thus supporting it during the bending operation.

Automatic Blank Feed for Forming Dies

A cam, a slide, and a magazine automate the feeding of flat blanks to a punch press. In Fig. 2, view X shows the blank and the channel to which it is formed by the operation.

The positions occupied by the components when the ram has completed its up stroke are shown in view Y. Cam *A*, operating with punch *B*, passes through an open area in slide *C*. Magazine *D* forms a bridge over the opposite end of the slide. There is a second open area in the slide, which is a nest for the lowermost blank in the magazine.

As the ram descends, the slide, with a blank in its nest, is first pulled in rapidly until the blank is aligned with the punch T and die E. Continued descent of the ram does not further move the slide, and the punch forms the blank, pushing it out of the bottom of the die, as in view Z.

The magazine has a spring-loaded movable bottom F which keeps the blanks from falling out. On the down stroke, the spring forces the bottom under the magazine as the slide moves into the press. On the up stroke, the slide displaces the bottom, and a new blank drops into the nest.

Because the work is ejected below the die and because of the presence of the cam, the press must have a stroke greater than would ordinarily be required. Both the slide end and the movable bottom should be lower in height than the thickness of the blank. Conversely, the bridge clearance of the magazine should be greater than the thickness of the blank. This is necessary for the smooth operation of the device.

Tube-slitting Punches Work Out-of-Phase

A press operation involving two punches that function out-of-phase produces a long slit required in the wall of tubing. The work is shown in view X of Fig. 3. Tubing diameter is 3/4 inch, with a wall thickness of 0.050 inch. The dimensions of the slit are 3/32 by 2 inches.

View Y shows the press setup. Tubing A rests on support B, the top of which is concave to conform to the diameter of the tubing. Since the piercing pressure is applied at an angle, it is necessary to restrain the tubing from moving axially. This is done by two toe dogs C.

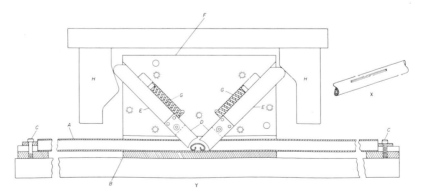

Fig. 3.   To prevent the punches D from interfering with each other, the profiles of the cams H cause the punches to advance and retract independently.

The punches *D* are replaceable blades, doweled and screwed to blocks *E*. These blocks slide in channels toed in at an angle of 45 degrees in tool body *F*. Compression springs *G* keep the blocks normally raised.

Bearing against the end of each block and mounted on the bottom of the die shoe is a cam *H*. As can be seen from their differing profiles, the two cams prevent the punches from interfering with each other.

When the press ram descends, the left-hand punch advances first, piercing the tubing, then retracts slightly. Further descent of the ram advances the right-hand punch, and the slit is completed. As the ram moves up, the right-hand punch retracts first. The left-hand punch next re-enters the slit, and then it, too, retracts. When the tubing is removed from the press, the slug drops out of one end of the work.

A satisfactory slit is produced, but because there is no supporting die within the tubing, some distortion is inevitable. To eliminate this, a proposed alternate design has the punches toed-in in a horizontal plane, with a spring-loaded semi-circular pad lowered over the top of the tubing.

Rotary Broach Designed to Shave a Part in a Punch Press

The piece shown in the detail view in the upper part of Fig. 4 is an adding rack used in one of the business machines made by the International Business Machines Corporation. The stamping is produced by blanking and piercing in a punch press, after which the hook-like ear on the side of the stamping away from the rack teeth is shaved by a special rotary broach in a subsequent press operation. The shaving operation is necessary in order to produce the required smoothness of face and to hold the 1.775-inch dimension to limits of ±0.002 inch.

The rotary broach is mounted in a die set, as illustrated in Fig. 4, and used in a punch press. Thus, this precise job can be accomplished in a press with a 2 1/2-inch stroke, rather than having to use a conventional broaching machine. The die provides a means of locating and clamping the workpiece, as well as actuating the rotary broach. It is opened by the press action, and is loaded and unloaded by hand, the workpiece being clamped and unclamped automatically.

As the die is closed, a clamping lever *A*, Fig. 4, is swung downward by the rod *B*, which is actuated by the air-cylinder plunger *C*. The plunger is forced into the cylinder as the ram of the press moves downward. There is also a spring plunger *D*, Fig. 5, which presses against the workpiece when the die closes. The workpiece is located in the die by pins that fit into the outer ends of the slots in the piece. The location is such that the face cut by the broach is at the desired angle with respect to the straight edge of the work.

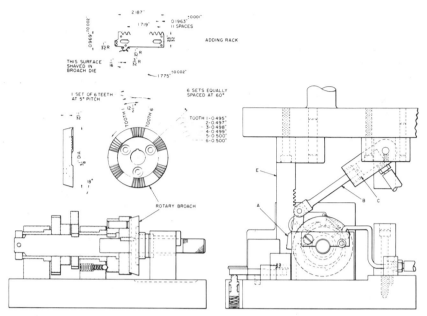

Fig. 4. Rotary broaching die designed to be used in a punch press for shaving a portion of the adding rack, illustrated at the upper left. Motion of the broach, shown in detail beneath the workpiece, is imparted by a ratchet and pawl actuated by a pinion turned by a rack.

Because a circular broach having six groups of six teeth, each with blank spaces between groups, is used, it is not necessary to give the broach a rocking motion or to move the workpiece away from the broach between cuts. These groups of teeth are on one face of the broach, 60 degrees apart. The spaces between the groups clear the work as soon as the last two teeth, which are of the same height, have made their cut. The first four teeth in each group make cuts 0.001 or 0.002 inch deep.

To actuate the broach, use is made of a rack E attached to the upper die member and meshing with a pinion in the lower member, the pinion being on the same shaft as a ratchet that is engaged by a pawl or latch. When the latch engages one of the ratchet teeth on the down stroke of the upper die member, the broach is turned 60 degrees to make its cut. On the up stroke, the latch is disengaged and the broach does not turn, being held by a spring loaded detent.

Since the broach is turned only 60 degrees at each down stroke of the press ram, every set of broach teeth is used only once in six strokes of the die. The broach is on a separate shaft that is in line with the pinion-shaft, and turns with the latter when the ratchet latch is engaged. As the broach teeth are on the face of the cutter, they produce a flat face on the end of the hook. With the workpiece securely clamped and the broach so supported that it cannot move axially, it is quite easy to keep the required dimension within the specified limits.

The controlling factor in the speed of production is the time required to unload and reload the die and trip the press, as the shaving action is almost instantaneous. A guard which has been removed in the view of the press set-up in Fig. 5, is placed in front of the die to protect the operator, and safety is further promoted by using a trip that requires both hands to be placed on trip-buttons.

Fig. 5. Rotary broaching die set up in a press, the safety guard being removed for clarity. The workpiece is positioned on pins and is held by a rocking clamp and a spring plunger.

Combined Bushing and Clamp for Hydraulic Piercing

In piercing operations performed on a mechanical or hydraulic press it is frequently necessary to clamp the workpiece against the die-block and keep it in the clamped position until the punch has been withdrawn. Normally, a spring stripper plate is employed to accomplish this clamping action, but when using a hydraulic piercing unit with limited space between the unit and the workpiece, another arrangement must be used.

Figure 6 shows a piercing unit in which a floating bushing *A* mounted in a liner bushing *B* is essentially a dual-purpose bushing. Not only do the combined parts align piercing punch *C* for the operation, but they also clamp the workpiece *D* against die-block *E*.

Liner bushing *B* is held in position by a dog-point set-screw *F*. The floating bushing unit is forced against the workpiece when contact occurs between a large neoprene washer *G* mounted on the piercing punch and the flange of bushing *A*.

To obtain the proper stripping effect, distance *X* is less than distance *Y*, which means that the bushing clamps the piece before the punch contacts the piece. After the piercing operation has been completed and the punch has been withdrawn, the compressed neoprene washer *H* under

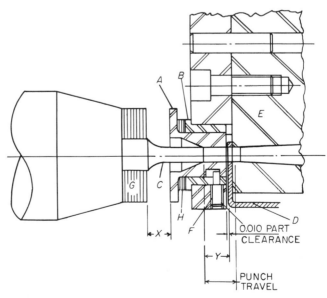

Fig. 6. Combination floating bushing and clamp for guiding punch and securely holding work during punch withdrawal.

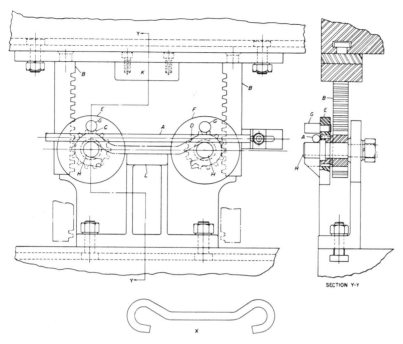

SECTION Y-Y

Fig. 7. Ingenious tooling designed for forming eyelet rods, such as seen at X, from
flat stock.

the bushing flange forces the bushing away from the piece. This action
restores the 0.010-inch clearance shown in the drawing. The same
design principle may also be used in drill jigs when conditions warrant
its adoption.

Bending Device for Double-Eyelet Rod

Tooling designed for simultaneously bending an eyelet at each end of
a short metal rod is illustrated in Fig. 7. The workpiece is seen at X.
Incidentally, the eyelets are closed somewhat more than shown by using
a hand tool when the rod is installed on agricultural equipment.

The tooling was designed for use on a hydraulic press, this type of
equipment being especially desirable because racks for actuating the
mechanism could be conveniently mounted on the plunger of the press.
Another advantage derived from this type of equipment is that the press
stroke is adjustable and can be set to the exact amount required for the
racks to operate pinions that actuate the tooling.

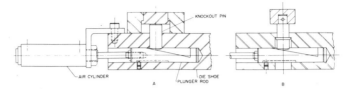

Fig. 8. In view A, the knockout pin operates under the control of the air cylinder, which is timed with the upstroke of the press. Where the knockout pin is not contained above the die, it is provided with a shoulder and moves in a counterbored hole, as seen in view B.

In an operation, the straight bar is laid on the tooling as shown at *A*. When the press plunger descends, it lowers racks *B* which are attached to the press ram. These racks revolve pinions *C* and *D*, causing pinion *C* to turn in the counterclockwise direction, and pinion *D* to rotate clockwise.

The revolving pinions cause plates *E* and *F* to turn and to carry pins *G* downward around the respective circles. These pins bend the ends of the rod around stubs *H* to form the eyelets. Block *K* bends down the center of the rod until it rests on top of anvil *L*, as indicated by dot-and-dash lines.

Minor changes in the tooling could be made for closing the eyelets more completely. In such a case, it would be necessary to increase the stroke of the press. This inexpensive tooling has saved a great deal of time and hand work in making eyelet rods.

Air Actuates Knockout

When running high-production progressive dies, it is often necessary to knock the part out at the final station. This is particularly true if the part is bent. In many instances it is desirable, and sometimes imperative, to have the knockout arrangement deactivated while the die is closing and to operate and retract during the upstroke of the press. In such cases the knockout pad cannot be spring-loaded, so that it will not be in an up position during that part of the cycle when it is wanted out of the way.

Shown in view A (Fig. 8), is a simple yet positive method of obtaining such action. A hole is bored in the bottom die-shoe to a slip fit for a plunger. This plunger, made of drill rod and hardened, is machined at an angle to contact the knockout pin and is tapped in one end for attachment to the connecting-rod of a small air cylinder. A small slot is milled lengthwise for a dog-point set-screw which prevents the rod from turning during operation.

The air cylinder itself is a commercially available item, either double-acting or single-acting with a spring return. It is mounted on an angle-bracket which is bolted to the shoe. In use, the cylinder is actuated on the upstroke by a cam on the press crankshaft.

An alternate arrangement that can be employed in case the knockout pin is not contained is illustrated in view B. By this method a hole is counterbored from the bottom of the shoe and a shoulder turned on the pin to limit its travel. A mechanism similar to this can be attached to the top of the die-shoe if sufficient space is available, or if the design of the die lends itself to such a mounting.

Tooling for Right-Angle Bends in a Power Press

The forming of right-angle bends in a punch press often gives unreliable results due to a number of tool, machine, and material variables. Through experience, toolmakers have found techniques by which they overcome forming obstacles. Major elements which have contributed to the lack of uniform results in the forming of right-angle bends are as follows:

1. Variations in hardness and thickness of material being formed.

2. Accuracy requirements of the squaring limits on a 90-degree bend. (More difficult to hold in the forming of short-leg heights.)

3. Troubles arising from large radii on an inside bend, especially true when the inside radius is greater than the thickness of the material.

Figure 9 shows a forming block with an angle of 4 to 5 degrees. A slight variation in the angle may be more efficient, based on actual tests.

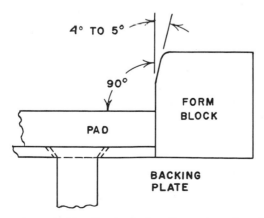

Fig. 9. This arrangement of bending tools gives the most consistently accurate results.

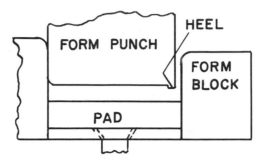

Fig. 10.   Forming with a punch having this kind of heel is risky with thin gages.

This becomes a starting angle which tends to perform its forming operation with a smooth rolling action rather than the usual sudden shock. Parts formed in this manner are more uniform dimensionally, even with the commercial variation of material thicknesses.

Some designs call for the use of a heel on the forming punch, Fig. 10. This technique however, is not practical on thin materials because of possible fractures.

Another design, Fig. 11, may have the shape of the punch member less than 90 degrees (to overcome spring-back), and a spring pressure-pad positioned at a corresponding angle. But this again may give mixed results, especially where variations in stock thickness must be considered.

While the methods described in Figs. 10 and 11 have been accepted and are being used by the toolmaking trade, experience proves that the method described in Fig. 9 is the most successful in special or extreme cases. It not only produces a more uniform product, free from form-block marks, but it is more economical to maintain.

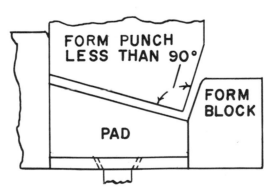

Fig. 11.   Variations in stock-thickness angular errors in the product with this under-90-degree angle punch, which is typical of tools designed to control spring-back.

Bushings in Thin-Wall Sections Increase Die Life

Tool and die repair shops have the obligation of prolonging die life. In simple terms, a punch and die set is a costly investment, and its conservation can result in worthwhile savings. Manufacturers are keenly interested in any means by which production, quality, quantity, and costs may be improved.

In repairing dies, it is often required to improve output without risking other production essentials. The following case is an example of how increased output per unit of die cost can be built into a die member.

Punch and die design standards will often dictate the need for rigidity in cutting areas. This is especially true when heavy thicknesses of material are being cut, imposing many kinds of risk. The section shown in Fig. 12 is of a die which cuts 3/32-inch thick low-carbon steel sheet stock used in the manufacture of mounting plates for telephone apparatus.

The initial design called for solid die construction because of the apparent weakness which would be created by the use of die bushings of normal size. This was accepted for years without objection. However, experience gained in tool operation revealed a short die life because of sheared holes caused by punch misalignment due to partial cuts and slugs. Die sections had to be ground beyond the sheared area in sharpening. This often meant a reduction of die thickness of 0.020 inch and a corresponding reduction of die life. Ordinarily, sharpening would be accomplished with a 0.007-inch reduction in die thickness. The die was was actually being ground away at approximately three times the normal rate.

Such waste could not be tolerated. After analyzing the situation, the use of bushings were considered for die conservation. This called for

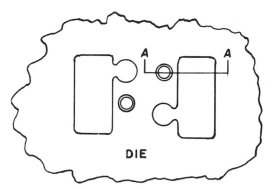

DIE

Fig. 12.   To improve the life of this die, bushings were added in spite of the thin-walled sections between the holes and the die openings.

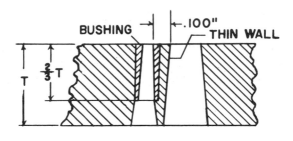

SEC A.A.

Fig. 13. Wall thicknesses were maintained at a minimum, and bushing depth was held to two-thirds that of the die.

several deviations from current standard design procedures, and the following changes were adopted (Fig. 13):

1. Die wall thickness was held to the minimum sufficient to withstand die pressures.
2. Depth of the bushing was limited in order to obtain maximum die strength.
3. Die-bushing wall thickness was held to a minimum.
4. Extreme care was used in fitting the die bushing in order to minimize strain and prevent breakage of die wall.

It is rather difficult for a designer to recommend thinner proportions than standard practice would dictate. Generally, the design is criticized, and if breakage occurs in the area of the thin section, it may be attributed to inexperience on the part of the toolmaker. However, properly applied skills will, in some instances, permit the use of die bushings where the minimum space between the perforated opening and larger die openings is approximately 0.100 inch, as in this case.

Whereas valuable die life was formerly ground away when holes sheared, use of the bushing has increased die life approximately 3 to 1. When excessively sheared, the bushing is replaced or shim sleeves are employed. The same practice can be applied to similar tools. In this instance, the bushings had a wall thickness of 0.050 inch and a length two-thirds the thickness of the die. Press-fit allowance on the diameter was kept under 0.0003 inch.

Punch- and Die-Holders Designed for Quick Tool Changes

One style of standard punch and die-button sets has a small depression, or drill spot, in the periphery of the members. This depression accommodates a set-screw which secures the tool to its holder. If a kick press has to be set up frequently for piercing different sizes or shapes of holes, the punches and die-buttons can be rapidly interchanged by removing the set-screws and using the holders illustrated in Fig. 14. The only restriction is that the outside diameters of all punches and dies be identical.

Punch-holder *A* and die-holder *B* are bored to a close slip fit with their respective members. In each holder there is a plunger *C* having a detent at its end to engage the depression. The other end of the plungers forms a link with a spring-loaded lever *D*. The punch and die are snapped into position. A movement of the levers release them. For compactness, the punch-holder lever is vertical, and the die-holder lever is horizontal.

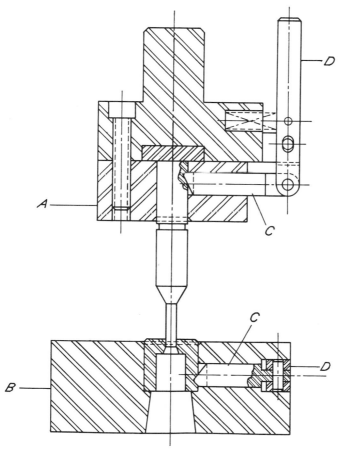

Fig. 14.   Replacing the set-screws of conventional holders are lever-operated plungers *C* having detents at their ends.

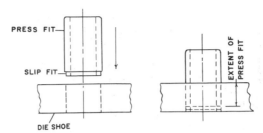

Fig. 15. Conventional method of assembling a guide bushing in a die-shoe.

Accurate Bushing Assembly

Precise squareness between bushings and die-shoes is one of the most difficult conditions to achieve in shop assembly operations. The usual procedure in pressing a bushing into a die-shoe is indicated by the diagrams in Fig. 15. A shoulder of relatively short length is provided on one end of the bushing (shown exaggerated at the left), this shoulder being a slip fit in the die-shoe. The remaining length of the bushing is made a press fit in the shoe. If a bushing of this design is squarely started into the die-shoe hole, the assembly can be completely satisfactory. However, if it is cocked even slightly, the toolmaker is in trouble.

The method of assembly shown in Fig. 16 is much more satisfactory because close squareness can be obtained with less caution on the part of the toolmaker. With this method, the entire thickness of the die-shoe is used as a lead for the long bushing shoulder. The bushing is assembled from the side of the die-shoe opposite to that from which assembly is started in the method illustrated in Fig. 15. This suggested method may also be used for assembling guide pins.

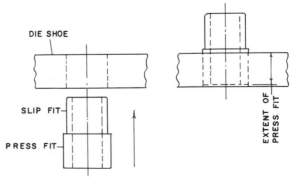

Fig. 16. Bushing assembly method which insures true squareness of components.

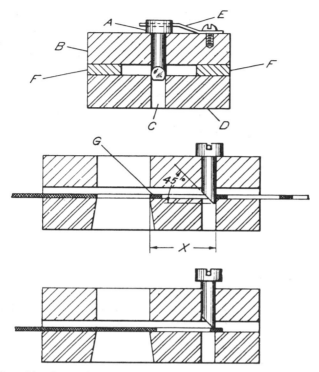

Fig. 17.   One side of the bridge left by each blank serves as a locating surface for the
following blank.

Semi-automatic Stop for Blanking Dies

Figure 17 illustrates a simple yet efficient design of a semiautomatic stop for a blanking die. A surprisingly high rate of production can be obtained with the aid of this device, since it is inherently quick-acting and accurate. The upper view, a cross-section from the front of the press, shows a stop-pin A fitted in a hole in stripper B. For the stop-pin, a short piece of drill rod can be used, being long enough to extend slightly into a clearance hole C in body D when the press is in the open position.

A slot in the head of the stop-pin accommodates a spring E, which serves to keep the stop-pin down in the stripper. The stop-pin is located midway between the stock guides F and to the rear of the die cavity, as may be seen in the center and lower views from the side of the press. Distance X from the back of the die cavity to the back of the stop-pin is made equal to the pitch of the work strip. (Pitch, in this instance, is the distance measured over the bridge G, between identical points on two adjacent blank areas.)

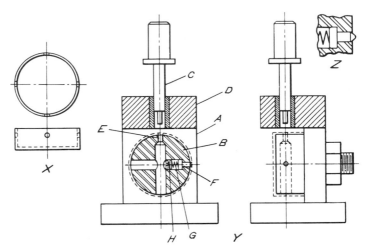

Fig. 18. Pin *F* engages each hole in sequence after the cup is pierced and rotated 90 degrees.

An important element in the design of the stop-pin is the chamfer on its tip. This chamfer is in the direction of the die cavity and is at approximately a 45-degree angle from the horizontal. The length of the chamfer must be greater than the thickness of the stock that is being blanked.

In operation, the stock is inserted between the stock guides, and the first blank is cut. (The front to rear position of the stock for the first blank is gaged visually.) The stock is then advanced, and when it strikes the chamfer on the stop-pin, the stop-pin is momentarily raised, as shown in the lower view.

Advancing the stock further, until the bridge passes the clearance hole *C*, permits the spring *E* to force the stop-pin down into the area left by the first blank. Then the stock can be pulled a short distance toward the front of the press until this movement is arrested by the contact of the front side of the bridge with the rear of the stop-pin, as in the center view, and the next blank can be cut. This procedure is repeated for all subsequent blanks.

Automatic Locating Fixture for Sequential Piercing Operation

A simple punch-press fixture with an integral indexing pin permits the rapid piercing of four small holes in the wall of a cup. The holes, spaced 90 degrees apart, are produced in sequence without taking the cup from the fixture.

In Fig. 18, the cup is shown at X, with a cross-sectional and side view of the fixture at Y. Bolted to the upright block A of the fixture is a stub arbor B, the outside diameter of which is made to a sliding fit with the inside diameter of the cup.

The punch C is of conventional design, and operates within a bushed guide plate D, entering the piercing hole E on the vertical center line of the arbor in the plane of the punch. A shouldered pin F is on the horizontal center line. It is shown enlarged in view Z.

The pin has a conical point which projects slightly from the arbor periphery, in which position it is normally kept by the pressure of a spring G against the pin shoulder. Set-screw H takes the thrust of the spring.

When loading a cup, the pin is forced in. After the first hole is punched, the cup is manually rotated 90 degrees toward the pin, which is now free to move out and engage the hole. When the cup is rotated after the second hole is punched, the pin is automatically forced in, moving out to engage this hole when it is opposite the pin. The cup is then properly located for the piercing of the third hole. The fourth hole is produced in an identical fashion.

Telescoping Bushing that Increases Stripper-Bolt Travel

When designing large forming, blanking, or drawing dies, it is sometimes impossible to obtain enough stripper-bolt travel. The extent of this movement is confined to the thickness of the die-shoe. One way to overcome the condition is to place risers on the bottom of the die-shoe. However, this method is not only costly but often cannot be accomplished because of the limited opening of the press. The solution to this problem is both simple and inexpensive.

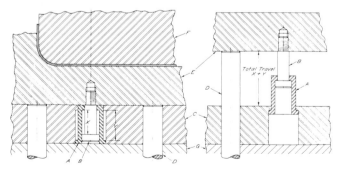

Fig. 19. Details of die provided with stripper-bolt that telescopes into bushing to permit additional travel.

In Fig. 19, the telescoping bushing *A* is counterbored to receive the stripper-bolt *B*. Die-shoe *C*, in turn, is counterbored to be a slip fit in the telescoping bushing. In the open position, shown at the right, pressure-pins *D* support the die *E* to which the stripper-bolts are attached. As the punch *F* descends, the telescoping bushing drops down on bolster plate *G*, and the stripper-bolt retracts into the bushing. This is the closed position, seen at the left. The total travel can be computed by adding distance *Y*, from the bushing shoulder to the bottom of the counterbore in the die-shoe, to distance *X*, from the shoulder of the stripper-bolt to the bottom of the counterbore in the bushing, when the die is closed.

If still additional travel is necessary, the bushing *A* can be made to telescope into a second, larger bushing. However, this method is seldom used as most dies do not need an extreme amount of travel.

Punching Small Holes in Heavy Stock

It is universally considered impossible to punch holes in mild steel sheet much smaller in diameter than the sheet thickness. If punching of smaller holes is attempted, punch wear will be too great and punch breakage will occur too frequently. Small holes can, however, be punched in heavy stock by performing the operation in three steps instead of in one step, as is customary.

Stepped punching takes advantage of a known fact that for severing a slug from a metal sheet, it is not necessary for the punch to go through the metal. It is enough if there is more or less pronounced penetration. The heavier the stock, the less the percentage of penetration that is needed. After the slug has been made a shaving punch is used to bring the hole to the desired size. The required pressure for shaving is much less than that for punching.

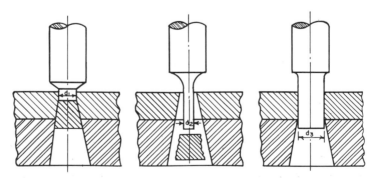

Fig. 20.   Three steps to be followed successively in punching small holes in sheet metal.

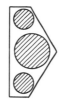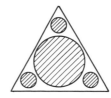

Fig. 21.   Diagrams illustrating possible layout of holes drilled prior to punching irregu-
lar-shaped holes.

For the first step, use is made of a high-grade shouldered punch with a short, strong cutting point as seen at the left in Fig. 20. This punch penetrates a short way into the material, sufficiently to sever a slug. The shortness of the cutting point d_1 insures necessary punch strength. A second punch with a small point diameter d_2 is next used to merely push the slug out of the hole, as seen by the central view of Fig. 20. A third step consists of shaving away the superfluous material from the hole, whose tapered, ragged walls are unsuitable for functional purposes. The hole is shaved to size and shape by a punch of d_3 diameter, as seen at the right.

Stepped punching may be performed not only on round holes, but also in case of irregular shapes. On high-production jobs, stepped punching may be incorporated in progressive dies, with the successive steps located strategically.

On short-run jobs the comparatively small quantities of punched holes may not justify the expense of three-step tools. In such cases, holes may be predrilled in sheet-metal parts instead of prepiercing and then shaved to the required outline. Figure 21 shows three diagrams that indicate how a few holes may be drilled prior to shaving away the excess metal. The round preliminary holes should be laid out by the designer so that the metal to be shaved away is fairly uniform around the whole contour.

Design of Trimming Punches to Eliminate Bending

The workpiece shown in the upper view of Fig. 22 originally was blanked with a rather complicated punch, but is now made much more easily and cheaply with a trimming and shearing die, as indicated diagrammatically.

Trimming punches, however, have a drawback in that the entire outside contour is not used in cutting, and there is, therefore, a tendency for the punch to bend away from the cutting force, as shown at A.

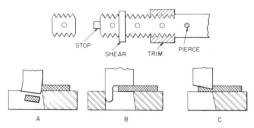

Fig. 22. In operation, trimming punches have a tendency to bend as seen at *A*. This can be overcome as indicated by views *B* and *C*.

This produces stresses in the punch, and also affects the quality of the work. To avoid this trouble, trimming punches and dies are often made wider than is strictly necessary to provide the punch with a heel, as seen at *B*, that enters the die before the punch touches the stock. In this way, the die supports the punch, so that it will not bend during the trimming operation.

Another way to solve this problem is to bevel the face of the punch, as shown at *C*. The wedging action of the bevel forces the punch toward the stock, thus counteracting its tendency to bend away from the stock.

The exact amount of the bevel angle is important, but can be determined only by practical trials. It depends upon such factors as thickness and hardness of the stock and die clearance. The best method is to begin with 5 or 6 degrees and increase or decrease the angle according to the results obtained.

Simple Methods of Mounting Small Punches and Bushings

Small punches are usually made from drill rod and hardened. They generally must have some kind of head so that they can be removed from the holes punched by them. The simplest and most frequently employed method for retaining punches in standard tools consists of peening over the head as shown at *A* in Fig. 23. This method has a drawback in that such punches are easily deformed, since the peening is done by hand, and, therefore, is not perfectly uniform. Often the head breaks during use because of inadequate head-treatment. In addition, peened heads are inadequate for comparatively high stripping pressures.

Square-headed punches *B* are usually employed in high-quality tools. Such punches are turned from stock of larger diameter than the punch body, leaving an integral head of sufficient size. The chief disadvantages of this design are economical in character: an increase in the costs of labor and material.

The solution of this problem consists of employing straight punches with an adequately designed fastening and stripping method. In cases where comparatively small stripping forces are involved, a simple design would be a straight piece of drill rod having a laterally milled flat for a set-screw, as seen at *C*.

For larger stripping pressures, an inexpensive design consists of providing an adequate circular groove toward the end of the punch to receive a split ring which can be made from steel piano wire. This method, shown at *D*, has been found excellent in actual practice. Of course, all of the methods described for punches can also be used for fastening die bushings. The four basic designs applied to bushings are illustrated at *E*, *F*, *G*, and *H*.

Stripper Spring Adjustment for Compound Dies

When a stationary punch incorporating a piercing die serves as the blanking punch member of a compound die, the resharpening of the punch necessitates readjustment of the height of the corresponding movable stripper. This adjustment must be carefully made in order to maintain the original surface relationships among the various members of the die.

If the adjustment is effected by shortening the holding screws, as is usually the case, the space allowed for the helical springs that back up the stripper will eventually become too small to permit the springs to operate properly. Actually, helical springs should not be compressed more than one-quarter to one-third of their free length. The repeated shortening of the holding screws is also troublesome.

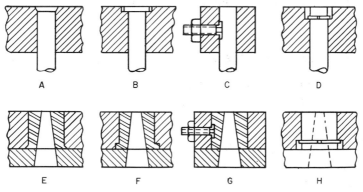

Fig. 23.   Straight diameter punches or bushings with a set-screw *C*, *G* or split-ring *D*, *H* mounting are economical and effective.

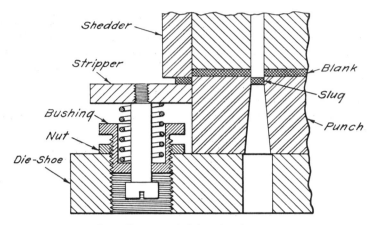

Fig. 24. Arrangement for adjustment height of stripper after regrinding blanking punch. The threaded bushing is screwed into the die-shoe a distance equal to the amount the punch is shortened. Thus, the space for the coil springs is not reduced, as the holding screws are not shortened.

To overcome these difficulties, the arrangement illustrated in Fig. 24 was developed. This permits easy, quick regulation of the stripper height with micrometer sensitivity, and maintains a space of constant height for the helical springs. A threaded bushing fits into a tapped hole in the die-shoe, and is clamped in position after setting by means of a lock-nut. The bushing fills the double purpose of housing the coil spring and limiting the upward travel of the holding screw.

Each time the punch is sharpened, the bushing is screwed into the die-shoe, a distance equivalent to the amount the punch is shortened by the regrinding operation. Thus the holding screw is not shortened, the space for the coil springs is not altered, and the original height relationship between stripper and punch is maintained.

Die with Cams that Operate Punches in Opposite Directions

If a hole is punched in a part before bending and is located close to the bending zone, it will be distorted during the operation. For this reason, it is frequently preferable to pierce holes in stampings after they have been formed.

The U-shaped stamping shown at X, Fig. 25, presented an interesting tooling problem. The blank is cut from 0.4-inch wide cold-rolled steel strip stock by means of a cut-off die that incorporates punches for the two holes in the bottom of the part. The problem was to produce the two elongated holes. Drilling was impractical, and the proximity of the holes to the bends made it necessary to punch them after forming. A conventional die, punching from the outside, would not serve the purpose since the hole would be located too near the edge of the required internal die-plate. Therefore it was desirable to design a punch and die that would pierce from the inside of the legs.

The solution was found by designing a novel die that would function within the comparatively small space available between the legs of the stamping. The part is located in a nest formed by grooves in two stripper plates *A*, Fig. 25, and locked in place by a quick-acting toggle clamp *B*. As the press is actuated, cam-holders *C*, Fig. 26, which are attached to the ram, descend and the punching cams *D* contact matching surfaces of the two punch-holders *E*, as seen at the right-hand side of Fig. 27. The punch-holders, together with the punches *F*, are forced outward to punch the holes in workpiece *G*. It will be seen that contact surfaces of the cams and punch-holders are machined at an angle.

Two stripper cams *H* wedge between the punch-holders and the stripper plates on the return stroke, forcing the punches inward to their

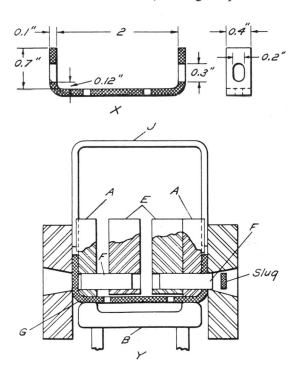

Fig. 25.   Plan view showing left-hand side of die prior to punching, and right-hand side in the act of punching. The cams are not shown.

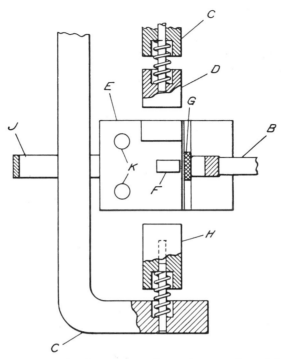

Fig. 26. Diagram illustrating how compression springs are situated between punch-holders *C* and cams *D* and *H* for safety.

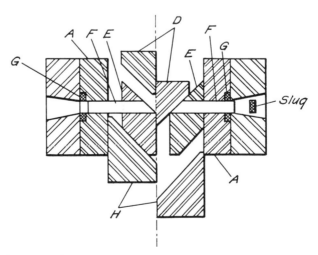

Fig. 27. Sectional view showing punching cam in operation at right and stripping cam in operation at left.

original position as seen at the left-hand side of Fig. 27. It was necessary to provide a positive-acting stripping device, as the limited space prohibited automatic stripping by means of compression springs. At the end of the return stroke, the completed parts are ejected by means of a manually actuated device *J* that pushes against the ends of the legs.

The cams are mounted free in their holders so as to avoid the necessity of fine adjustment of the press-ram stroke. A compression spring provides a small amount of overtravel of the cam-holder after each phase of the cycle. Alignment of the punches is maintained through a sliding fit of the punch in the guide hole and by two alignment pins *K*, shown in Fig. 26, located between the punch-holders and stripper plates.

Ram-operated Knockout for Second-Operation Dies

Although positive-acting, automatic ejectors are generally best, there are some instances where it is preferable to use some special design for such devices. In second-operation dies, it is possible to employ the usually inactive phase of the operation cycle for this purpose.

A certain U-shaped component — for an important functional reason — required absolutely square corners. In accordance with current practice, the workpieces were first shaped in a standard channel type forming die with rounded edges, as shown at X in Fig. 28 — the inside

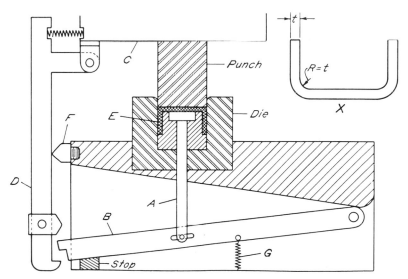

Fig. 28.   Ejector pin *A* retracts fully to allow easy loading of die. Part as received for this sizing operation is shown at X.

radius being equal to stock thickness. The bottom was then flattened by re-striking in a sizing die.

The common design for such second-operation dies is with the part inverted. It is easy to construct a die that will squeeze the bottom into correct shape and at the same time obtain the exact height for the two legs, with the work in this position. But with this type of die, ejecting becomes a problem since the squeezed workpiece tends to stick firmly in the die cavity.

An automatic knockout of a standard design failed. Since the part must be completely lifted from the die, it was quite difficult to find sufficient space for the long and heavy spring required to produce the necessary travel and high ejecting pressure. Also, the upper rest position of the ejector was too high to allow an easy introduction of the following workpiece into the proper position in the die.

A change-over from the spring-actuated ejector to one operated by the press ram, as illustrated, solved the problem. Here, knockout pin A is actuated by the pivoted horizontal lever B. On the upward stroke of the punch-holder C, the lever is engaged by latch D, which carries one end up until workpiece E is ejected. Just before reaching the highest position of the ram, the latch is disengaged from the lever by means of the tripper F. The ejector pin then retracts into the die, aided by a suitable spring G. This allows the next workpiece to be positioned in the die cavity with minimum effort.

Semi-automatic Feeding Device for Second-Operation Dies

Second-operation dies employed for piercing, shaving, or trimming often have feed-slides to transport the blank to the work area. The same device also provides a means of removing the finished piece from the die. These slides may be manually operated or cam-operated by means of press-ram action. A simple but efficient cam-operated die-feeding mechanism that provides automatic workpiece ejection is illustrated in Fig. 29.

When the press ram and punch A are at the top of their strokes, as seen in the two upper views of the diagram, feed-slide B is outside the working area of die C. At the same time, an auxiliary slide D, mounted at right angles to the feed-slide, covers the workpiece escape hole E.

With this arrangement, a blank dropped into the recess F in the feed-slide is pushed horizontally by cam G into the piercing position and pierced on the down stroke of the ram. Just prior to the actual piercing, cam H activates the auxiliary slide to uncover escape hole E. The two

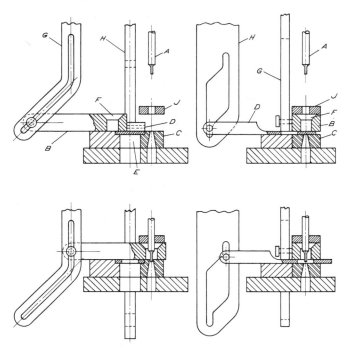

Fig. 29.   On upward stroke of press ram, feed-slide *B* allows workpiece (not shown) to escape through hole *E* previously uncovered by auxiliary slide *D*.

lower views show the relative positions of the two slides with the punch fully descended.

On the upward stroke of the press ram, the part is stripped from the punch by stripper *J* and is then carried from the die by the feed-slide. During this movement the workpiece passes over and through the escape hole in the auxiliary slide. A conveniently placed chute or bin receives the finished work. The auxiliary slide returns to its starting position at the end of the upward stroke, thus covering the workpiece escape hole.

Multiple Notching of Inclined Slots

Four uniformly spaced inclined slots had to be made in a small component destined for use in a lawn sprinkler. Milling of the slots proved too slow and expensive, so it was decided to stamp them. The press tool built for the operation is illustrated in Fig. 30.

The blank *A*, a flat disc, is placed in a suitable nest (not shown), on the die-plate *B*, and the press is actuated. While descending, the punch-

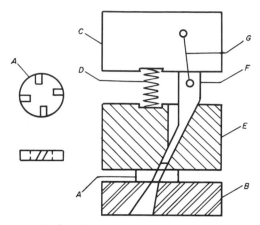

Fig. 30.   Die for piercing four inclined slots in a small disc.

holder *C* actuates a couple of strong compression springs *D* that cause a movable stripper and blank-holder *E* to contact the blank and hold it firmly. Further descent of the punch-holder causes four guided punches *F*, spaced at 90 degrees around the periphery of the stripper, to simultaneously produce the four inclined slots.

As the punch-holder rises, the four punches are lifted by means of short pieces of steel cable *G*. A rigid connection between the punch-holder and the punches would not be practical because of the required movement of the latter. After stripping the punches, the movable stripper is lifted and the finished workpiece is manually removed from the tool.

Expanding Mandrel for Coil Stock

A mandrel for coil-stock feeders features three radially adjustable arms which compensate for differences in coil hole sizes among the various suppliers. Because it has three arms instead of the customary four, the device can exert a more effective grip when a hole is out of round.

The mandrel handles coils weighing up to 1 ton, with hole diameters from 9 to 17 inches. It is automatically self-centering, and is tightened and loosened by turning a handwheel. In use, the coil is placed on the mandrel, which is adjusted beforehand to approximate size. Then the mandrel is tightened, expanding the arm diameter about 1/4 inch with each turn of the handwheel.

A drag brake prevents overriding and helps produce a smooth feeding action. The mandrel always rotates in a direction which tends, due to the drag, to tighten it against the coil.

In Fig. 31, cylinder *A* is supported and held in place by two roller bearings *B* and thrust bearing *C*, and is free to rotate around shaft *D*. The entire external surface of the cylinder is threaded: One half is a right-hand thread, and the other half is a left-hand thread. Each half is engaged by a nut *E*. These nuts move toward or away from each other when, by means of handwheel *F*, the cylinder is caused to rotate independently of the stationary shaft.

Each of the three gripping arms *G* (only one is shown) is connected by two toggles *H* to the nuts. In addition, there is pinned to the center of one toggle of each arm a restraining link *J*. The other end of the link is pinned to a stud on ring *K*, which is confined axially by the end of the cylinder and plug *L* but is free to rotate during adjustment of the mandrel. Purpose of the link is to assure that the arm remain parallel with the cylinder and in the same vertical plane. It is, of course, necessary that the link swing on the stud at the intersection of the lines describing the motion of points *M* and *N* of the toggle.

Drag is produced on the rotating members by the thrust of collar *P* against plug *L*. The collar is keyed to shaft *D*, and the amount of thrust it exerts on the plug can be adjusted by turning the spring-loaded nut *Q* in either direction.

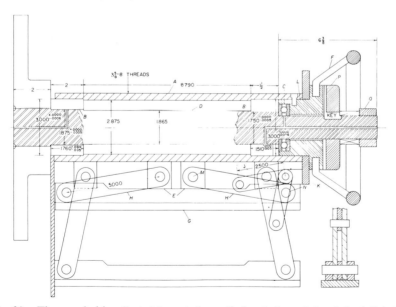

Fig. 31.   The mandrel is adjusted by rotating cylinder *A*, through handwheel *F*, independently of shaft *D*. This causes nuts *E* to move laterally, and thus, through toggles *H*, causes the gripping arms *G* to move radially.

To tighten or loosen the mandrel, the expanding linkage must be prevented from rotating with the handwheel. The weight of the coil is sufficient to prevent this movement. If the mandrel has to be adjusted before a coil is positioned on it, a restraining hand can be placed on one of the arms.

Overforming Lanced Tabs in Sheet Metal

Lanced tabs are generally formed in sheet metal to a 90-degree bending angle. If the bending angle must be less, the operation requires somewhat more care, as the press stroke must be carefully adjusted.

In the case of lanced tabs with bending angles greater than 90 degrees it is not possible to use simple, ordinary tools. The dies must have special features.

For short-run jobs it is recommended that the job be performed in two separate operations. First, a right-angle lanced tab is formed in the standard manner and then, with a separate bending tool, the tab is overformed to the desired angle. In the case of mass-production jobs the complete tab forming should be effected in a single operation. If only a small amount of overforming is required, the punch bottom for slitting the three sides of the tab and then bending it should be shaped as shown at *A* in Fig. 32. Then the end of the tab will be bent first with respect to the rest of the tab, and as the operation progresses, the rest of the tab will be bent to a right angle and its end will be overformed.

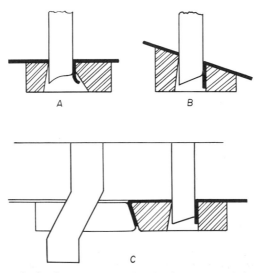

Fig. 32.   Three methods of overforming tabs that have been slit in sheet-metal parts.

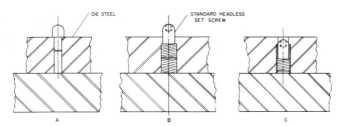

Fig. 33. Diagrams illustrating conventional and improved types of pilot-pins for dies.

The die-plate opening must be provided with adequate clearance for the preformed tab.

If the tab must be straight and the bending angle is large, a standard punch as shown at *B* is preferable. The die-plate is inclined (to the desired angle) with respect to the punch axis. In such cases, the sheet metal has a tendency to slip downward so that it is necessary to provide stops and clamps or blank-holders. Roughening the top surface of the die-plate will help.

In progressive dies it is not convenient to work with inclined die-plates. The operation should be performed in two successive steps. First, a right-angle tab should be formed and then overformed in the second station. The overforming may be effected by slides, pivoted or hinged arms, or hinged levers. An arrangement with cam-actuated slides is shown at *C*.

Improved Design for Die Pilot-Pins

Pilot or locating pins used on various type of dies are usually turned from bar stock to the shape shown at *A* in Fig. 33. The pilot-pin is made with a short shank. This shank is pressed directly into a hole in the die-block. Whenever the time comes for sharpening the die, the pilot-pins must be removed. Frequent driving of the pins in and out of dies eventually causes them to be loose in their holes. There is then the possibility of the pins being picked up by the work unnoticed by the operator. Serious damage can result to the die or press in such an instance as well as injury to the press operator.

An improved yet simple method of making pilot-pins is shown at *B*. Use is made of a standard headless set-screw turned or ground down on the back end to the desired pilot diameter. The threaded remainder of the screw is entered into a hole drilled and tapped from the underside of the die steel. Such a pilot-pin can be retracted as shown at *C* whenever the die must be sharpened.

Actually, a pilot-pin of the improved design is stronger than the conventional pin because it does not have a small diameter shank readily subject to breakage. Also, because all up-and-down adjustment of the pin is controlled from the top, it is unnecessary to remove the die-block or even turn it over in order to retract the pilot-pins. Top dies can also be fitted with pilot-pins of this type.

CHAPTER 14

Unusual Dies

Die with Floating Blankholder

Designed to produce a small spring plunger, the die shown in Fig. 1 has a floating blankholder, and a combination spring stop and work ejector.

At each stroke of the ram a complete part is produced. Wire is automatically fed from a straightening device into the die and passes freely through a hardened bushing that functions as a stationary shear blade for cutoff. The movable shear blade, which is attached to the punch holder, has a half-round groove sized to fit the wire. Two small opposed punches form the shoulder on the part.

In operation, the wire is introduced through the shearing bushing into a groove in the blankholder. The correct length is gaged by a spring-loaded stop. The stop is adjustable and, being spring-loaded, is pushed back by the wire, which is lengthened during the shoulder-forming operation. It returns to its original position when the die opens.

Once the wire is located the press is tripped and the descending upper blankholder comes into contact with the wire and holds it; the wire is sheared; the blankholder lowers it below the shearing bushing; and the shoulder is formed between the two punches. The upper blankholder is set to push the floating blankholder down only the distance required to correctly form the wire.

When the ram ascends, the part is lifted by the lower blankholder. The right end of the part strikes the lower portion of the shearing bushing, the part is knocked over, and slides down through a chute. The heavy compression spring of the stop helps to eject the part. Strokes of the upper and lower blankholders are limited by bolts located transversely to the wire feed direction.

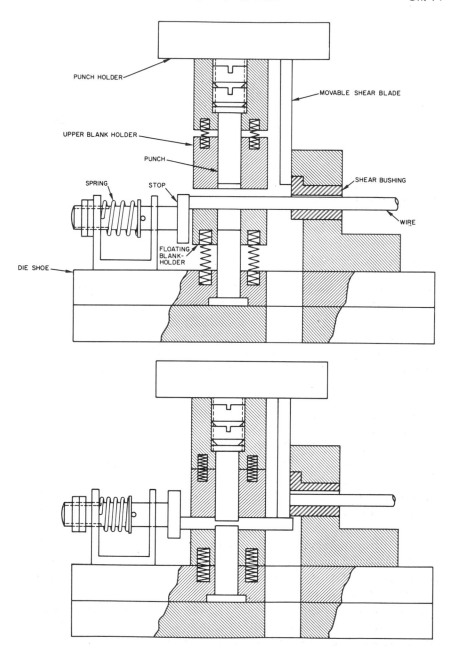

Fig. 1.   Die for producing small spring plunger has a floating blankholder.

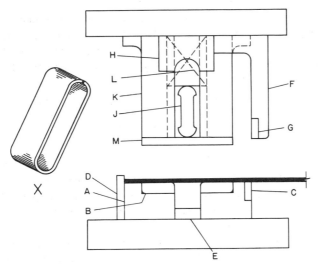

Fig. 2.   Workpiece, seen enlarged at X, is fabricated from strip stock in one operation
with the die shown. Horn J is in the fully extended position.

Die Has Floating Horn to Allow "Closing" of Flat Stock

Workpieces, originally made from strip stock in three separate operations, can be formed with one stroke of the press ram by a unique die. Previous operations, requiring three dies, included shearing to length, curling of ends, and bending of center.

The part shown at X in Fig. 2 is made from 0.060-inch thick hot-rolled steel that is first sheared to 1 1/4 inches in width and then fed through the die in 4- to 12-foot lengths. In Fig. 2, the die is illustrated in the open position with the stock in place. The lower member consists of two blocks A, with reversible and hardened members B at the bending corners; a reversible, hardened insert C at the shearing edge; and a stock stop D. A hardened bumper block E is located at the bottom of the die opening.

The upper half of the die supports the shearing leg F which has a hardened and reversible shearing insert G at the bottom. Closing block H is located at the top of the die above horn J which slides vertically in gibs K. Parts H and K are both hardened. Heavy springs L force the horn down against plate M at the bottom of the guides. The gib construction and arrangement for locating and holding the horn are illustrated in Fig. 3.

In operation, the stock is placed in the die and positioned against the stop. The press is then tripped, and as the ram descends, the stock is first sheared to length. Since the shear blade projects only 1/16 inch below the bottom of the horn, there is insufficient time for the stock to move out of place.

The momentum of the ram and the heavy springs acting against the horn serve to bend the blank into a U-shape, as shown at the top in Fig. 4. Continuing to descend, the horn forces the workpiece into the die opening until it strikes the bumper block *E*. Here the workpiece and horn stop, while the ram, in finishing its stroke, forces the closing block *H* against the blank to complete the operation. At the bottom in Fig. 4, the die is shown in the closed position. Due to spring-back, the finished part is readily removed from the horn either by hand or by mechanical knockouts.

The entire die is made of cold-rolled steel except for the horn, closing block, and the hardened bending and shearing inserts. By making these inserts reversible, the die need not be removed from the press to renew the edges. The gibs are covered with hard-bronze alloy wear plates (not indicated) and kept well lubricated. This die has successfully produced thousands of parts with minimum down time.

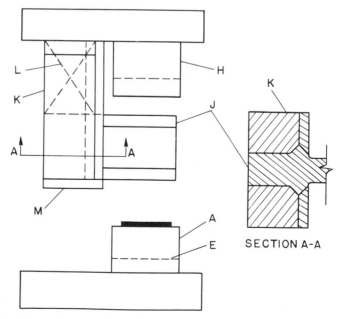

Fig. 3.   Relative positions of closing block *H*, horn *J*, and die-blocks *A* are more clearly shown in side view of the die. The horn slides within the gibs *K* (section A-A).

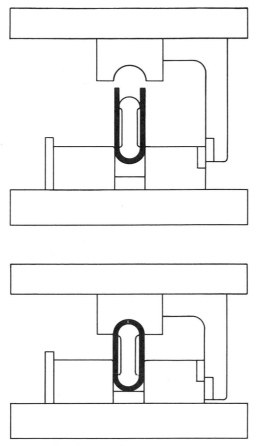

Fig. 4.   With the ram partially descended as seen at the top, the stock has been sheared and the center portion bent.   In the lower view, the ends are "closed" to complete the piece.

Split-Level Progressive Dies Blank and Deep-Draw a Shell

Parts handling and tool costs are slashed, and size change-overs are easier, using the novel cupping tools in Figs. 5 through 7. For the sake of simplicity, the shell that will be described here (*W*, Fig. 5) is blanked and cupped in three blows. However, it has proved practical to make cups requiring several draw strokes. Figure 5 is a schematic view of the tools from the front, and Fig. 6 is a view from the right, with the ram up. A side view, with the ram down, is shown in Fig. 7.

Block *A* is mounted on plate *B*, which is fastened to the bolster plate. Block *A* carries the blanking die *C*, the stripper *J*, and the first drawing

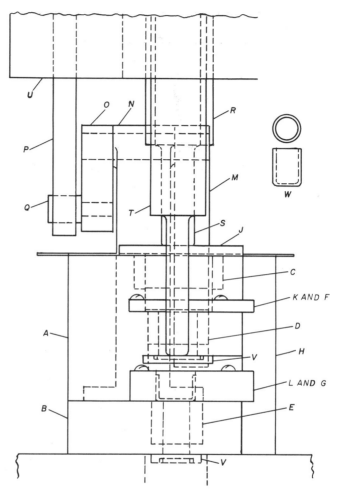

Fig. 5. Schematic front view of the blanking and cupping tools. A pivot reciprocates the slide carrying the draw die *D* and stripper *F-K* with it. *W* represents a typical workpiece.

die *D*. The second drawing die is carried in plate *B*. Block *A* is grooved to carry the slides *F* and *G*, which are tied together by two plates *I*. These plates are separated to support roller *X* between them. Slides *F* and *G* are retained by plate *H*, shown only in Fig. 5, and are provided on their outer ends with semicircular recesses, which on slide *F* are slightly greater in radius than the radius of the blank, and on slide *G*, slightly greater than the radius of the first drawn shell. Block *A* and plate *B* carry the blocks *K* and *L* with matching semicircular recesses so

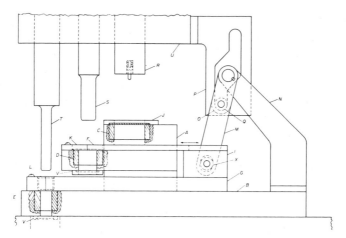

Fig. 6. Schematic view of the right-hand side of the split-level progressive tools shows the cam and lever mechanism *O-M-Q* which reciprocates draw slide *I*. At its extreme right-hand position stripper *F-K* catches a blank from die *C*. The first draw is produced by punch *S* in the position shown.

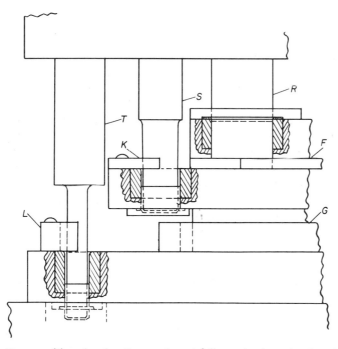

Fig. 7. Draw position showing the punches at full cup depth. *R* has knocked out a blank; *S* has made a first draw; and *T* has drawn the full cup. The ram descends further to eject blanks to the transfer.

that when slides *F* and *G* are in the forward position, as shown in Fig. 6, the blank and the first drawn shell are held in position for the next operation. Strippers *V* are provided beneath *D* and *E* for removing the shells from the punches. These strippers are discs, recessed to carry a commercial snap-ring of slightly smaller inside diameter than the outside diameters of the shells in their respective stages of manufacture. The shells are forced through the rings, making them open. The snap-rings contract after the shells have passed completely through, thus stripping the shells from the punches.

Punch-holder *U* carries the drawing punches *S* and *T*, the blanking punch *R* (which is equipped with a spring plunger), and cam-plate *P*. Bearing bracket *N*, mounted on plate *B*, supports a horizontal shaft which carries the arm *M*, keyed to it, which engages roller *X*. Arm *O* is keyed to the other end of the shaft, forming a bellcrank lever, and carries the roller *Q* which engages cam-plate *P* to transmit reciprocating motion to slides *F* and *G*. The contours of the cam surfaces provide a period of rest at the beginning and end of both the "up" and "down" strokes of the ram. Slides *F* and *G*, therefore, act as transfers. *P* carries blanks to *D*, and *G* carries cups to *E*.

Referring to Fig. 6, which shows the positions of the parts at the top of the stroke, the blank which has been produced on the second down stroke has been brought forward and is held in position for the first draw by block *K* and the end of slide *F*. Similarly, the shell that has been formed in die *D* on the previous down stroke is held in position for the last draw by block *K* and the end of slide *G*.

In Fig. 7, the ram is sketched as having descended sufficiently to return slides *F* and *G* to their rear positions. The roller *Q* (Fig. 6) is now at the beginning of the dwell period following the end of the down stroke. The movement of slides *F* and *G* does not begin until punches *S* and *T* have entered their respective dies sufficiently to prevent the shells from being drawn out of position by the rearward movement of the slides. In the position shown, the blank has been formed and the two shells have been drawn but not ejected. Continued downward movement of the ram will force the blank out of its die into the recess on the end of slide *F*. The first drawn shell will be forced out of its die into the recess on the end of slide *G*, and the completed shell will be ejected from its die, dropping through the bolster plate. On the upstroke, slides *F* and *G* will remain at rest until the punches have withdrawn from their respective dies far enough to offer no interference to the movement of the slides.

The material used is 0.040 inch thick. Theoretically, there is no limit to the thickness of the material, provided that the various sections can be proportioned to withstand the blanking and drawing pressures. This, in turn, is governed by the rated power of the press. Any metal with sufficient ductility to permit drawing without fracturing may be used. Shells of this type have been produced of low-carbon steel, brass, and aluminum. One of the determining factors is the depth of draw and the radius of the punch at the bottom of the shell.

The softest material that will satisfy the requirements of the product should be used. For steel, the ideal hardness is about 99 Brinell. Harder material may be suitable for shallower draws but requires a larger punch radius at the bottom.

Using carbon-steel die bushings, between 700,000 and 1,000,000 shells should be produced with a set of tools, though it will be necessary to occasionally sharpen the blanking punch and die. Using alloy tool steel, greater die life can be obtained. By the use of a semicarbide, die life may be increased three to four times.

In any drawing operation, die life is governed by two additional factors: proper lubrication and cleanliness of the material. A good grade of drawing compound, which must not be re-used without filtering, is essential. Bright finished stock, free of grit and shop dirt, is also essential.

Horn Die for Trimming, Knurling, and Grooving Drawn Parts

Punch press shops are often confronted with the problem of knurling, grooving, or trimming drawn metal cups or similar parts when lathes and other means of rotating the work are not available. In such cases, the work can be performed on an ordinary punch press equipped with a horn type die such as shown in Fig. 8.

The rack A, attached to the upper member of the die, actuates a pinion B on the down stroke of the press. The pinion is connected to an arbor C, which rotates the work when the press ram moves downward. A plunger D, also attached to the upper member, contacts a knurling or other tool holder E. The tool-holder is hinged to a support, (not shown), and is normally held away from the work by spring pressure. The spring F, acting on the plunger, applies pressure to the work for any depth of knurl or groove required.

An ejector collar G is mounted on work-arbor C for quick removal of work. Collar G is actuated by a bell-crank H connected to a foot-lever. When the work has a tendency to slip, instead of rotating with the arbor, an expanding arbor can be used. Cups having a wall thickness of only 1/32 inch have been trimmed quickly and efficiently with this type of die.

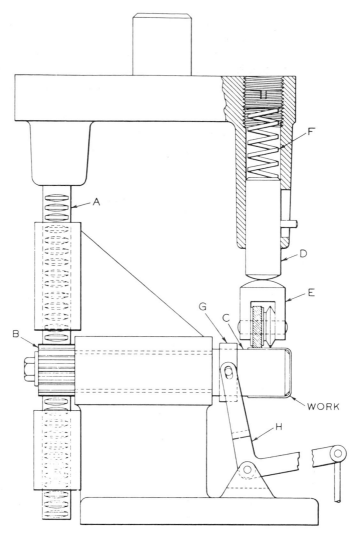

Fig. 8.   Horn die with rotating arbor designed for performing knurling, grooving, and trimming operations on drawn metal cups.

Cable-operated Die for Multiple Piercing

An unusual die employed for piercing six equally spaced holes in the periphery of a drawn aluminum cup is illustrated in Fig. 9. The six punches are simultaneously activated by an arrangement of sheaves and a stainless-steel cable. Although the die is designed for piercing, the principle can be adapted to other press operations.

In construction, a vertical plate A is secured to a baseplate B. A hollow cylindrical die C, machined on the periphery to accommodate the workpiece D, is attached to the vertical plate with dowels and cap-screws. The punches E are retained and guided in radial bores in another hollow cylindrical member F which is also secured to the vertical plate. Member F, being bored to receive the workpiece, is positioned concentric to the die. A small sheave G is mounted on a dowel pin H in a slot at the end of each punch. In each punch assembly, the ends of dowel pin H project from the sides of the punch to retain a short compression spring J. These springs serve to retract the punches from the workpiece. The drive cable K is wrapped around the sheaves, as shown, and each end is secured by a clamp L to an arm M attached to the upper die member N. As the ram descends, the cable pulls the punches radially inward toward the center of the die to pierce the cup

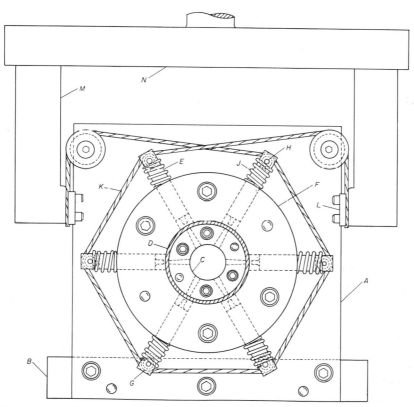

Fig. 9. Multiple-piercing die that is operated by a cable attached to the upper die member.

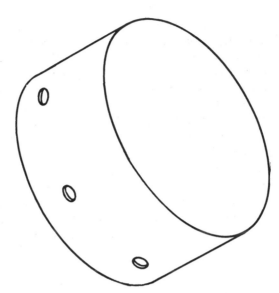

Fig. 10. Six radial holes are pierced in the periphery of this drawn aluminum cup by means of the die shown in Fig. 9.

(Fig. 10). A hole is bored in the vertical plate in line with the hole in die member *C* to facilitate removal of the slugs.

This die was used in a press having a 1 1/2-inch stroke. In a press having a longer stroke, each end of the cable could be arranged as seen in Fig. 11. Springs *P* hold the cable in contact with the sheaves while allowing some additional movement of the press ram. A smaller press may be used if the die is placed horizontally.

Novel Wedge Type Forming Die

Quite frequently an indentation of some particular contour has to be formed in an already drawn and pierced cup (Fig. 12). An indentation added to such a cup is shown in section X-X in Fig. 13. The difficulty generally encountered in the use of most wedge type form blocks for an operation of this type is that gaps are present when the blocks have been expanded to full diameter. However, with the unique design shown in Fig. 13, no such gaps exist at full expansion. This condition effects a more desirable indentation of the cup.

When the forming blocks *A* and *B* are placed in the cup, they are in the positions shown at the right in Fig. 13. A wedge block *C* is placed between blocks *A* and forced downward by the ram of either a punch press or an arbor press. This causes the blocks *A* to move outward, which, in turn, gives a similar movement to blocks *B*. When all four blocks have attained their maximum expansion, their outer edges are at their true diameter *D* and the indenting process has been completed.

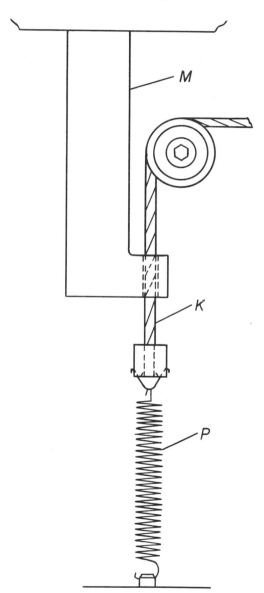

Fig. 11.   An alternate arrangement of cable *K* that can be used if press has a long stroke.

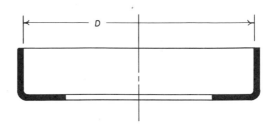

Fig. 12.   Drawn cup before forming on the periphery by the die shown in Fig. 13.

The final positions of the forming and wedge blocks and the finished indentation E, are seen in section X-X and in the left-hand plan view in Fig. 13. In this position no gaps exist between the blocks.

After indentation, the wedge block is removed first and then form blocks A and B are returned toward the center of the cup, as shown at the right in Fig. 13. The unloading clearances F and G are illustrated at the upper right, and clearance G, in section Y-Y. All remaining blocks are then removed from the cup. This design may be used whenever an accurate uninterrupted forming surface is desired.

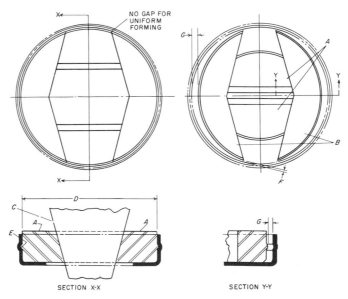

Fig. 13.   Forming blocks A and B are inserted and removed from the drawn cup when in the position shown at the right. Wedge block C is forced down by a press ram to expand the blocks to their final forming position (left) without leaving a gap. Wedge block C is not shown in the two upper views.

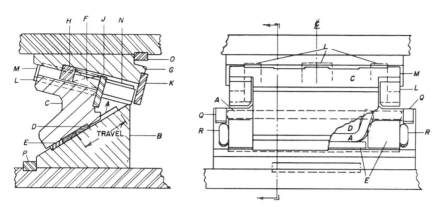

Fig. 14. Economical cam-action die which accepts interchangeable inserts for bending various shape parts.

Versatile Cam-Action Bending Die

It is the responsibility of the die designer to cut production costs wherever possible. One way this may be accomplished is to provide tooling that can produce a number of similar parts in basically the same die simply by changing a few die inserts. A cam-action die designed to accept changeable inserts so that parts of various shapes can be bent and formed is seen in Figs. 14 and 15. In addition to its versatility, this die is relatively easy and economical to build since no pressure-pads, stripper bolts, or heavy compression springs are required.

The workpiece A (Fig. 14) is placed between a fixed slide B and a sliding cam C, attached to the upper portion of the die. As the press ram descends, bending is accomplished by inserts D (secured to cam C)

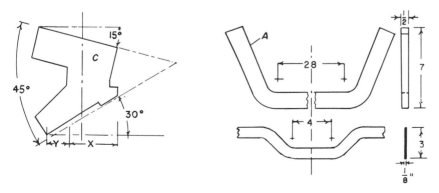

Fig. 15. Here shown are working angles for cam member C (left) and typical parts (right) that can be bent in the die.

and E (secured to slide B). The cam travel along slide B and the bending occur at an angle of 30 degrees to the base of the die (Fig. 15). Four medium-strength compression springs F are used to provide an equalizing force operating against cam C as it moves along a cam-holder slide G (and slide B), but only during the last 5/8 inch of the working stroke. Two springs F are boxed in on each side of cam-holder slide G by means of a retainer H (secured to the cam-holder slide) and a plate J mounted on cam C. Plate K, attached to member G, acts as a stop for cam C on its return stroke along cam-holder slide G during the upward movement of the press ram.

Most cam-action dies are designed so that the return stroke of the cam member is accomplished purely by spring action. Here, the return stroke is accomplished mostly by gravity, with a short initial push being given the cam by the equalizing springs F. Thrust plate L, guide plate M, and wear plate N are made in the usual manner. In this design, keys O and P are a must in order to keep fixed slide B and cam-holder slide G in place. Oil-grooves and supply lines are provided for proper lubrication of the cam-slides.

In Fig. 14, where the die is cut away, the workpiece A is seen after the press stoke has been completed. The dotted lines show the same part nested between gage members Q before the bending operation. Keys R help retain the outer tooling inserts E in place. A standardized arrangement can be made for easy interchangeable mounting of various bending inserts D and E. In this way, similar parts, such as those shown at the right in Fig. 15, can be formed in the die. The working angles for cam C are shown at the left in the same illustration. Work area X should be twice the free guide length Y.

This method of bending is particularly good for parts less than 3/16 inch thick. Workpieces of this type will be straight at the critical point of the bend, since the vertical thrust of the press ram holds them flat and forms them at the same time. Heavier stock is easy to keep straight.

One Die Makes Four Different Parts

Four different components of an adjustable slide assembly installed on printing machinery are made from one blanking and piercing die. Only a negligible amount of change-over is necessary in order to produce the several parts. These are shown at the top of Fig. 16. All parts are made from 14-gage (0.0747-inch thick) cold-rolled steel sheet, 2 inches wide.

Parts a and c are the same except that part a lacks a 3/16-inch hole. Both parts are formed to a 90-degree angle, while parts b and d are flat.

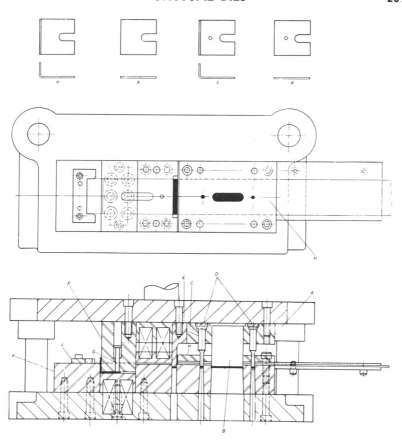

Fig. 16. Progressive die which can be applied with minimum change for producing four different parts from flat stock.

The die is constructed with a punch-holder pad A made of cold-rolled steel which is bolted and doweled to the top shoe. Slotting punch B is made twice as wide as the length of the slot in the parts so as to pierce a double length slot as indicated. Two standard piercing punches C are used for the small round holes in parts c and d. Two slides D, to which punches C are attached, are pulled out of the die to deactivate the punches when making parts a and b. Parting punch E is attached to the top shoe with screws extending from the top downward to facilitate removal for sharpening. The combination cutoff and forming punch F is backed up by a small spring-loaded plunger G to make certain that the part never hangs to the punch on the upstroke.

In starting a new strip of stock, the operator pushes the stock through channel guides slightly past the end of the positive stripper plate H. A

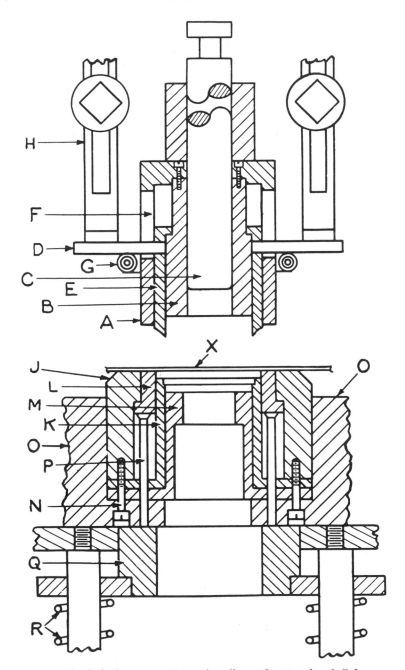

Fig. 17. Combined single- and double-action die produces a rim shell from excess
stock in the process of forming a drawn shell.

small portion is cut off the end of the strip and the two small round holes and a slot are then punched. The strip is next advanced to stop J, which is fastened to the top of form block K. Thereafter, every stroke of the press will produce one part of c and d or a and b if the piercing punches are inoperative. Punch F cuts the strip through the middle of the slot.

The dies is used on a tilted press so that the parts fall to the back in separate containers. A sheet-metal chute guides the formed and flat parts into separate containers.

Combined Single- and Double-Action Die Produces Two Shells in One Operation

Rim and drawn shells of the design seen at a and b, Fig. 18, are produced in one operation from the same strip of stock by using a combined single- and double-action die mounted on a double-action power press. Sorting is eliminated since the two types of shells are ejected from the die separately at the rate of 240 pieces a minute. The construction of this die is shown in Fig. 17. The press is equipped with an automatic reel type feeding device.

The top cross-section in Fig. 17 shows a single-action blanking and drawing rim punch A secured by countersunk screws to a double-action punch sleeve B. A plunger C that draws shell b is a sliding fit in sleeve B. Rods D are screwed into the rim ejector sleeve E and move freely in slots F. Coil spring G (under the rods) encircles sleeve A and provides the tension necessary to support the ejector sleeve. Brackets H are adjustable to required heights above the rods to function in ejecting the finished rim shell.

The rim drawing ring L in the lower cross-section has a sliding fit between blank cutter J and the rimlip shaping sleeve K. The shell drawing sleeve M is aligned with the punch sleeve B. Sufficient clearance is provided between the two sleeves to facilitate drawing shell b. Screws N secure the stationary parts J, K, and M to die base O.

The moving assembly upon which rim drawing ring L rides consists of pins P attached to a buffer Q which is supported by compression springs R and reacts downward during the drawing of rim a.

In operation, see Fig. 18, the stock X is blanked by punch A as it passes the upper edge of cutter J, the blank being supported by member L. At this time, because of the outward movement of punch A, ring L is depressed, thus compressing springs R through the pins P and buffer Q. Immediately, the lip of the rim shell is formed by sleeve K.

As the ram of the press descends, the double-action punch B shears the lip end of rim a and forms the shell blank Y. With the continued

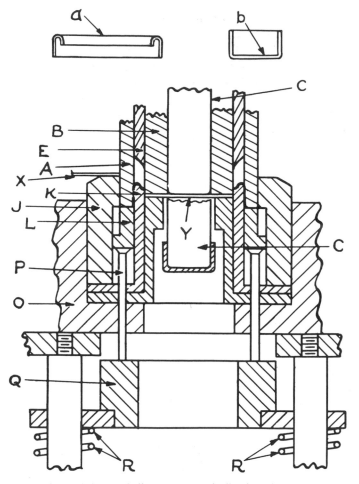

Fig. 18. Both rim and drawn shells are automatically ejected separately, eliminating the need for sorting them.

descent of plunger C, blank Y is drawn through sleeve M (Fig. 17) to form the shell b. On the upward stroke of the plunger C, the shell is stripped off the punch by a shoulder on member M and drops into a receptacle. Rim a is ejected at the same time by ejector E when rods D contact brackets H, Fig. 17. The rim shells are deposited in a box by an air blast coordinated with the operation of the press ram.

Plunger C, Fig. 18, is shown in two descending positions. It is regulated and timed so that it will draw shell b after punch B has sheared rim a and moved blank Y into position.

Double Compound Die

Two precise stampings were required for a product that was to be made in comparatively small quantities. In the interest of economy, the possibility of simultaneously stamping both parts in a single die was investigated. After some minor modifications to the outer contour of the smaller part, construction of such a die became feasible. The smaller part could then be made from the scrap from an aperture punched in the larger component, as illustrated in Fig. 19. In this way several advantages were obtained: no stock was required for the smaller stamping; both parts were produced at the same time and at essentially the same labor costs required for one part; and the cost of the single die was less than for separate tooling.

Construction of the die developed for the purpose is seen in two positions in Figs. 20 and 21. The outer contour of the large component is produced by the main die-plate *A* and the main punch *B*. The outer contour of the small component (and the large aperture of the large stamping) is produced by an auxiliary punch *C* operating against a die opening in the main punch. The three small holes in the large stamping are punched by secondary punches *D* also cutting against die openings in the main punch. Finally, the holes in the small stamping are punched

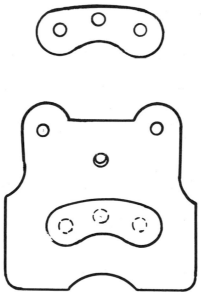

Fig. 19. A minor change in the original contour of the smaller part allowed it to be produced from the material from the aperture of the larger part.

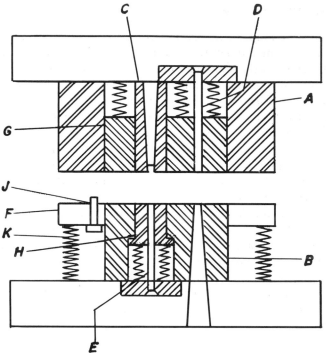

Fig. 20.   Double compound die used to produce both parts (Fig. 19) simultaneously, in a single press stroke.

by tertiary punches E against tertiary die openings in the auxiliary punch C.

Ejection of the stampings is effected in the following way: Stripper F separates the strip from the main punch B. The main shedder G ejects the large stamping from the interior of the main dieplate A, and the auxiliary shedder H ejects the small stamping from the interior of main punch B. These two shedders, at the same time, separate the two stampings from the small piercing punches D and E. The scrap from the large stamping falls from the die by gravity, whereas punchings from the small stamping are pushed upward and are carried by a sheet metal trough to the rear of the press. The stock-advance stop J is of the disappearing type, being simply a headed pin actuated by a flat spring K.

Blanked Parts Utilize Entire Strip Stock

The reverse F-shaped part, Fig. 22, coordinates good product and die design. It is made from 0.018-inch thick silicon steel strip and subsequently becomes an element of a lamination in a small transformer. The production of the part is of interest because, in being blanked from the strip, a second part of identical configuration is left as "waste." Thus, two parts are completed in each cycle of the press.

In addition, the parts utilize the entire strip stock. The lay-out, Fig. 23, shows the arrangement of the parts on the strip. There are no bridges left after blanking, the scrap being limited to small slugs of the four holes pierced in each part.

A diagram of the die setup is seen in Fig. 24. When a strip A is initially fed into the press, its end is located against a starting stop B while the required eight holes C in the first two parts are pierced. Thereafter, the end of the strip is located against a second stop D, so that while a part in work (beneath the punch E) and a "waste" part F are being formed, the holes in the next two parts are simultaneously pierced.

The part blanked by the punch is permitted to drop in the customary manner through the bottom of the die-bed. Provision for removing the waste part is made as follows: In advancing the strip against stop D, an ejector G is retracted against the pressure of a leaf spring H. When the punch clears the die on the up stroke of the press ram, the ejector pushes the waste part against the rear stock gage J, as in Fig. 25. The end of

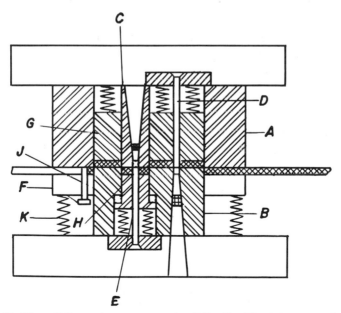

Fig. 21. Position of the various components of the die (Fig. 20) on completion of downward stroke of the press.

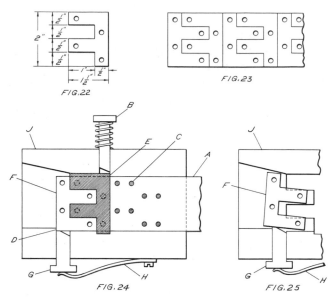

Fig. 22.　Except for the slugs of the pierced holes, this part utilizes the entire strip stock.
Fig. 23.　Adjacent parts interlock — the waste part on the left, the punched part on the right.　Fig. 24.　Before the punched part is severed, the ejector *G* is retracted.
Fig. 25.　The waste part is pushed out of the side of the die.

this gage is cut away, so that when the strip is again advanced, it pushes the waste part out of the die.

Trimming Die with Unique Features

A novel die was designed to perform a trimming operation used in producing the shallow drawn cup shown at the left in Fig. 26.　The part is made by blanking, drawing, piercing, and trimming in three operations.

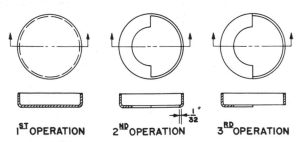

Fig. 26.　Part is first blanked and drawn into a shallow cup (left).　Lip remaining after piercing (center) is removed in third operation by trimming die.

A compound die is first used to blank the part from 20 gage steel strip and then draw it into a cup that is 1/2-inch deep and 2 inches in diameter on the inside. As the inner radius at the corner is only 1/16 inch, deep-drawing quality steel is used in order to eliminate tearing and popping.

The second operation, performed in a conventional die, pierces out most of the bottom of the cup (Fig. 26) but leaves a semicircular segment. Also remaining after this operation is a 1/32-inch lip extending halfway around the bottom edge of the cup wall. It is the trimming of this edge that led to the designing and building of an unusual die.

The troublesome lip is to be trimmed so that the lower edge at the open half of the cup is even with the inside surface of the portion of the bottom that remains. A burr-free lip without knife edges is required. Pinch trimming is therefore not permissible. Cam-operated trimming dies were also ruled out since two operations or complicated cycling of two or three cams in one die would be required.

The die developed for this trimming operation is shown in its final evolution in Fig. 27. Vertical clamping member A is made of cold-rolled steel and is mounted on the punch-holder. A hardened steel insert on the lower end provides a clamping and cam surface that can be replaced when worn. The shear blade B is made of hardened and ground oil-hardening tool steel. Flat-ground stock was used so that easy production of replacement blades is possible. Member A is offset from the shearing blade a distance equal to one metal thickness plus clearance. In this case the offset was 0.040 inch. Both these members are secured by screws and dowels to a cold-rolled steel supporting block that is not indicated in the

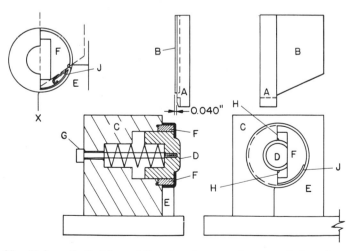

Fig. 27. Unique die that is used for trimming operation on a shallow-drawn cup.

illustrations. As shown in Fig. 27, the lower end of member *A* projects below the shear blade to clamp the part before the cut begins.

The die consists of a built-up body *C*, a sliding pilot *D*, and a shearing block *E*. For locating purposes, the stamping fits snugly over a turned and ground projection *F* on one end of the body. This projection, which also acts as a shear blade for part of the cut, is made a separate part to facilitate replacement. The pilot is spring-loaded and slides within the hardened and ground body of the die. A rectangular shank prevents rotation of the pilot which is retained within the cavity by bolt *G*. Locating edges *H* on the pilot bear against the edges of the straight section of the hole pierced into the bottom of the cup and position it in the vertical direction. When the die is open these locating surfaces project beyond the face of the body far enough to effect positive alignment of the part. The pilot is made of hardened and ground water-hardening tool steel. Although the pilot receives severe treatment during the operation of the die, it has not worn sufficiently to warrant replacement.

When the ram is tripped, the clamping member *A* pushes the sliding pilot completely out of the path of the shear blade by the time it has traveled about 1/4 inch into the work. During the first half of the trimming cut, the thrust of the shear blade tends to rotate the part in a clockwise direction. Turning is prevented by the now vertical edge of the bottom of the cup. This edge bears against the vertical portion of the shear blade, as seen at X in Fig. 27. Pressure on the work is quite great at this point, but buckling is prevented by the clamping member.

In trimming the lower half of the cup (as mounted), the shearing action takes place along the outer edge, and the shear block *E* is utilized. This is also shown at X in Fig. 27. A relief machined on the bottom of the projection *F* at *J* eliminates the necessity for a precision relationship between the part, the die body, and the edge of the lower shear block. This relief also facilitates loading and unloading of the die, prevents jams due to part tears and chips, and largely reduces wear on the die. This die has been in production for about three months and has proved quite economical to operate and maintain. Except for occasional heavy burrs forming where the cut shifts from *F* to *E*, quality of the trimmed edge has been consistently high.

Economically Designed Die for High Production of Small Clips

Illustrated in Fig. 28 is a small clip used extensively in the electrical equipment field. Since production of this item is extremely high, the die

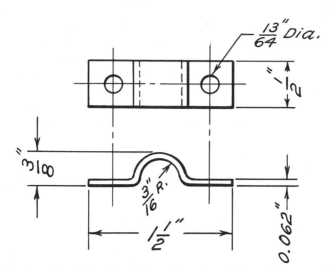

Fig. 28.   This clip is pierced, formed, and cut off in a single operation.

for it must be able to withstand the rigors of constant use and not be in the toolroom continually for repairs. Besides being accurate and rigid, the die must be simply constructed, so repairs can be speedily made when required.

A cross-sectional view of the die designed to produce the clip is shown in Fig. 29. Punch-holder pad *A*, made of cold-rolled steel, is screwed and doweled directly to the top shoe of the press, and carries the two hole-piercing punches *B* and the cut-off and form punch *C*. Block *D*, made of hardened tool steel, is mounted on the die-shoe and serves the dual purpose of being a stock stop and a back-up for punch *C*.

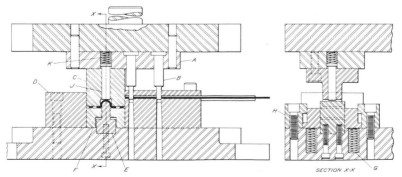

Fig. 29.   As can be seen in this cross-sectional view, all parts of the die are easily accessible when it becomes necessary to make repairs.

Form-block E is sunk in the die-shoe and held down with two cap-screws. Around this block is a stripper F which travels up and down over two springs G. Keepers H, which are screwed and doweled to the die-shoe, serve to restrict the travel of the stripper during the up stroke of the press ram.

In the top shoe is a hardened knock-out pin J that works against a spring K to eject the completed clip should it stick in punch C. All parts of the die are easily accessible.

A jet of compressed air synchronized with the up stroke of the press ram blows the clip into a bin at the rear of the machine, so that the operator's hands at all times can be a safe distance from the die.

Die Punches Four Notches at One Time

Resourceful die design has made it possible to produce four equally spaced indentations around the periphery of tubing in a single press stroke. The arrangement of the indentations in the work can be seen at X in Fig. 30.

End and elevation views of the die are shown at Y, the press ram A being at the top of its stroke. The tubing B is slipped over a horn C having slots in the areas where the indentations are to be made. This horn is fixed to a block D, which is the principal element of the die. Keeper plates E retain the block in alignment. The block is free to move vertically between the end-pieces F of the base G of the die. Fastened to the top of the block is a cross-piece H, suspended by bolts J from the punch-holder K.

An upper punch L is attached to the punch-holder, and a lower punch M is fixed in the base. The two side punches N slide horizontally in a

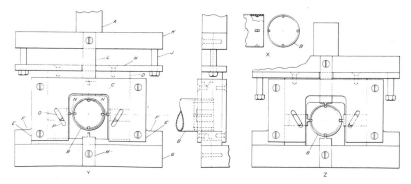

Fig. 30.   The cam action of the ribs O in the angular slots P causes the side punches to move toward the tubing on the down stroke of the press ram.

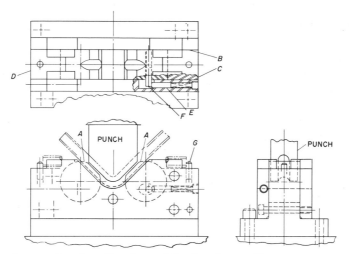

Fig. 31. Improved angle uniformity and freedom from galling are the advantages of designing tube-bending tools with rotating form-blocks *A*.

channel cut across block *D*. A rib *O* on the front of each side punch is a close fit in an angular slot *P* in its respective keeper plate. Since the keeper plates are stationary, vertical movement of the block produces a horizontal movement of the side punches.

On the down stroke of the ram, the block is free to descend behind the keeper plates, causing the side punches to move toward the tubing. When these punches abut the outside of the tubing, the movement of the block is arrested and the ram continues its descent independently until the punch-holder contacts the cross-piece. The ram and block then continue down as a unit, the side punches starting to indent the tubing. Shortly thereafter, the upper and lower punches come into contact with the tubing and start their indentations.

When the ram has reached the bottom of its stroke, as seen at Z, the block has descended until it bears on the base, and all indentations have been completed. Since the velocity of the side punches is lower than that of the ram, the side punches start to operate earlier so that all indentations are completed simultaneously.

Dies Bend Tubing Without Galling

Quality in bending seamless steel tubing comprises getting accuracy in the bend, smoothness of contour, and keeping the surface free from scratches and galling by the dies. The dies illustrated in Fig. 31 were designed for forming a 90-degree bend in a punch press.

To prevent excessive marring of the surface, the dies were built with the lower member consisting basically of cylindrical pivoting form-blocks *A*. Rounded grooves are ground in the blocks to cradle the largest diameter of tubing to be formed. The blocks have bearings in side-plates *B* and *C*, which are supported by plate *D*.

The punch has a groove identical to that in the form-block. Spring-loaded plungers *E*, on both blocks, bearing against pins *F*, return the blocks to the loading position at the end of each press stroke. When the dies are in open position for loading, both blocks have turned so that the grooves face upward in horizontal alignment. Rotation is stopped by pin *F* striking the wall of its clearance pocket in plate *C*.

Holder *D* is made wide enough to provide additional bearing surface at the top and at the ends of the form-blocks. Because side-plate *C* has a stop surface for pin *F*, it is doweled to holder plate *D*. But plate *B* needs no dowels. The same screw fastening holds both side-plates in position. Gage pins *G*, at each end of the die, position the parts at loading. Depth stop-blocks, not shown, are mounted on the die set to speed setup and control the angle formed.

While the cost of rotary forming dies is possibly somewhat greater than pad type dies, the bent tubing is always acceptable as it comes from the dies. This set of dies was designed for low costs in manufacture and maintenance. The high quality of product cuts scrap loss.

CHAPTER 15

Special Purpose Tooling

Synchronized Quad Vernier Dividing Head

As many as 3,577,772 divisions can be obtained from a universal dividing head having a total of only 179 index control positions. The mechanism, which is patented, is unique in that it is gearless, employing, instead, three discs adjacent and concentric to each other. Hole-circles are accurately located on the discs to form two vernier-matched pairs of stop positions, so proportioned that the output of one pair serves as the vernier for the other pair, in this way producing millions of divisions. The center disc, called the *control annulus*, if of a differential nature, except that there is no reverse rotation.

One of the problems of dividing has always been that of distributing the leftover digits of uneven divisions. This drawback has been eliminated almost entirely by the extremely large common denominator that the control annulus helps to produce. The infinitesimal size of the divisions possible delays decimals of leftover digits from accumulating so rapidly that they must be spread often to maintain working tolerance.

A drawing of the dividing head is shown in Figs. 1 and 2. The three discs referred to are: back disc A, control annulus B, and front disc C. The headstock D supports the hollow spindle E on which is secured a chuck or faceplate (not shown). The spindle is feathered by key F to the front disc, which is free to turn, as is the control annulus. The back disc, however, is fixed to the headstock casting by screws and dowels, and cannot turn.

Both back and front discs have single hole-circles, G and H, respectively. In the control annulus, there are two hole-circles — circle J, equal in diameter to circle G; and circle K, equal in diameter to circle H. Plunger L locks the control annulus with the back disc, and plunger M

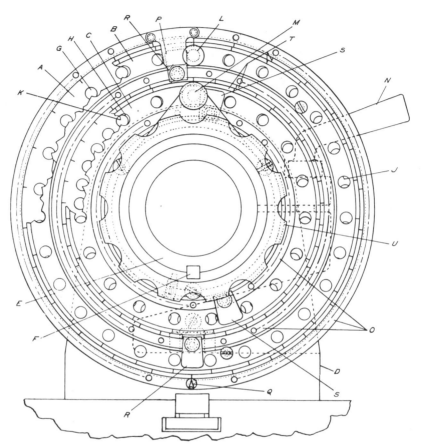

Fig. 1.   End view of synchronized quad vernier dividing head.

locks the front disc with the control annulus. When the head is in use, strain is kept from the plungers by locking the spindle directly to the headstock with the clamp *N*.

The 179 index control positions are provided for by the following hole-circle combination: back disc, 61 holes; control annulus, 31 holes (outer circle) and 43 holes (inner circle); and front disc, 44 holes. For clarity, a reduced number of holes is represented in each of the various circles in the drawing.

Three pairs of sector arms, used conventionally, indicate hole positions forward for the plungers. The arms are secured, by knurled screws, to a ring *O* of tapped holes located in a groove near the periphery of each disc. Arm *P* and pointer *Q* span the back disc; arms *R*, the control annulus; and arms *S*, the front disc. A change of setting is made by moving the arms to different tapped holes in their respective rings. The number of tapped holes in each ring corresponds to the number of stop positions.

Other features of the dividing head are two sets of vernier lines T — the quad vernier — to assist in setting the index holes to coincidence and a fluted wheel U for turning the spindle. While the device is of the "direct" type in that no gears are required, it can, if desired, be adapted to function through a conventional gear-reduction train.

The outer, 31 hole-circle of the control annulus is the vernier mate to the 61 hole-circle of the back disc. The difference fractionally across any adjacent 2 of the 61 spaces and 1 of the 31 spaces is 1/1891. This is the fractional forward movement of the control annulus when plunger L is transferred from one side to the other of the two slightly different angles — $2/61 - 1/31 = 1/1891$.

The inner, 43 hole-circle of the control annulus is the vernier mate to the 44 hole-circle of the front disc. The difference fractionally across 1 of the 43 spaces and 1 of the 44 spaces is 1/1892, the apparent reverse movement made by the front disc and spindle relative to the control annulus. This fractional rotation takes place when plunger M is transferred from one side to the other of the two slightly different angles — $1/43 - 1/44 = 1/1892$.

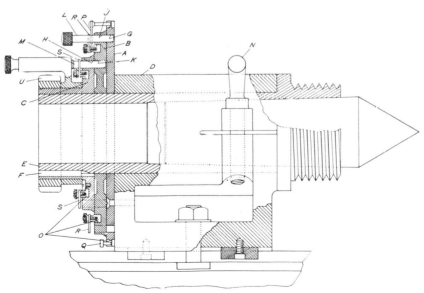

Fig. 2. Side view of synchronized quad vernier dividing head. Three discs of hole-circles acting as verniers on each other replace the gear train of the conventional head.

Problem I: Divide the angle 1/3,577,772

A possible way of obtaining this division would be, first, to reverse the front disc 1 hole (1/1892) relative to the control annulus by means of plunger *M*; and, secondly, to advance both the front disc and the control annulus one hole (1/1891) relative to the back disc, by means of plunger *L*. The difference between the advance and reverse (1/1891 − 1/1892) is 1/3,577,772, the net gain in rotation of the front disc and the spindle.

Another sequence of movement is preferred, however, since it avoids reverse rotation. When using it, both plungers still contact the same index holes as before. In this instance, plunger *M* is released from the control annulus and advanced 43 holes, with the aid of the appropriate sector-arm pair, around the front disc. Plunger *L* is then released and advanced 1 hole in the 31 hole-circle of the control annulus. Next, by pressing down and forward on the plunger, it aligns itself with the next hole in the back disc. Finally, plunger *M* is pressed down and forward until it aligns itself with the next hole in the control annulus. The division is now completed.

There are three basic angular measurement sizes in descending order, like degrees, minutes, and seconds: 1/61, 1/1891, and 1/3,577,772. As stated previously, the value 1/3,577,772 represents the difference between the secondary angle 1/1891 and an arbitrary angle 1/1892. Thus, computations must include the digit number of 1/3,577,772 "micro" units and the digit number of 1/1891 "secondary" units. In other words, for each 1/3,577,772 unit represented, one digit must be added to the number of digits representing the 1/1891 units.

Problem II: Displace the angle 1895/3,577,772

This angle is equal to the sum of 1/1891 and 3/3,577,772. There is a total of four digits, and the setting consists of moving plunger *L* 4 holes in the 31 hole-circle of the control annulus and moving plunger *M* 41 holes (44 − 3 = 41) in the 44 hole-circle of the front disc.

Problem III: Divide the prime number 157

Dividing 1891 by 157 gives a quotient of 12 secondary units and a remainder of 7 secondary units each equal to 1892 micro spaces, or a total of 13,244 micro spaces. Then, dividing 13,244 by 157 gives a quotient of 84, the micro-unit displacement, plus 56 leftover micro-unit digits. The 84 micro units are displaced by 1 hole on the 43 hole-circle of the control annulus and 40 holes on the 44 hole-circle of the front disc (84/44 = 1 plus 40 remainder). The sector arms *S* for the front disc are set to span 4 (44 − 40 = 4) holes according to the principle established in Problem I.

To obtain the displacement on the 61 hole-circle of the back disc and the 31 hole-circle of the control annulus, the 84 digits of micro units are added to the 12 digits of secondary units, for a total of 96 digits. These are displaced by 3 holes on the 61 hole-circle of the back disc and 3 holes on the 31 hole-circle of the control annulus (96/31 = 3 plus 3 remainder). The sector arm P and pointer Q for the back disc are set to span 3 holes, and sector arms R for the control annulus are also set to span 3 holes.

One of the main advantages of the 3,577,772-division capacity of the dividing head can be appreciated from the manner in which the 56 leftover digits are spread. The usual 400,000-division capacity head will require 56 corrections to spread a 56-unit leftover. But with each division of the new head only one-ninth the size, the 56 digits are spread in groups of 9, using only 6 corrections (6 × 9 = 54). After each of the 26 (157/6 = 26) divisions, an addition of 9 (9/3,577,772) micro divisions is made.

The control positions can be stopped in coincidence by means other than mechanical with plungers as described. They can be stopped by visual comparison, or by electromagnetic or electronic positioning devices. The head operates either clockwise or counter-clockwise with equal facility.

Novel File Support Applied to Machine

In machine-filing an opening in a die-block, the opening sides had to be square with the die-block base within close limits. The file had a tendency to deflect as pressure was applied, with the result that the die opening was being filed with tapered sides.

To remedy this undesirable condition, the toolmaker doing the job mounted a fiberglas roller A (see Fig. 3) in a steel holder B and applied it to support the back smooth surface of the file. The roller revolves on a hardened dowel-pin. A spacer C was placed between the roller holder and the filing machine table to permit clamping the assembly to the table by means of a C-clamp, as shown at D.

With this arrangement, the file was adequately supported as it oscillated up and down. A true vertical filing action resulted.

Keyslotting Tool for Extra-large Castings

In a shop, it is frequently necessary to cut keyways in castings so large that a conventional keyseater is impractical for the job. For work of

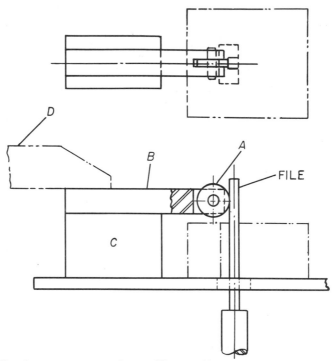

Fig. 3.   Simple arrangement used on a filing machine to obtain a truly vertical well-supported and straight filing action.

this character a special tool was designed that is applied on radial drilling machines. A circular keyseating cutter is rotated by the drilling-machine spindle, and the entire unit is fed down through the work by conventional means.

Referring to Fig. 4, body A is turned from AISI 1020 steel, to a diameter of 4.749 inches. In the case of the particular tool shown, body diameter is about 0.003 inch smaller than the hole in the casting. The body B is kept from rotating during operation by a hardened tool-steel guide. C. The one-piece shaft turns in bronze bushings D and E. Thrust is taken by ball bearings F and G.

The keyway for which this tool is built is 0.750 inch wide by 0.375 inch deep. Cutter H, made of hardened and ground high-speed steel, is carried on a tool-steel shaft. The shaft is so located in the body that the cutter protrudes beyond the body the distance necessary to cut a slot of the correct depth. The cutter has sixteen teeth machined to a 15-degree helix angle. The cutter teeth themselves mesh with a worm J, for driving purposes, as shown.

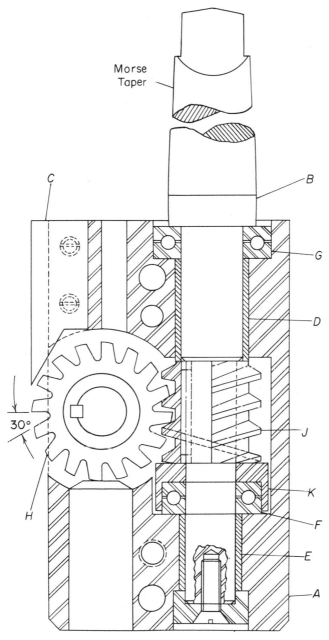

Morse
Taper

30°

Fig. 4. Worm *J* on shaft *B* is turned by a radial drill to drive milling cutter *H* as though it were a worm-wheel. The cutter will cut a keyslot in big bores of castings too large or too cumbersome to be handled by standard keyslotting machinery.

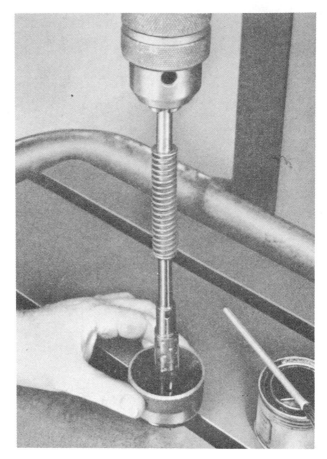

Fig. 5. Alignment of a lapping tool used in a drill press can be improved by this floating drive.

The worm is also made of high-speed steel and hardened and ground. It has a two-start thread and drives the cutter from the back of the teeth. Cap K prevents chips from entering the lower thrust bearing and transmits end thrust from the worm to the lower ball bearing. Infeed thrust is taken by the top ball bearing. Before hardening, the worm and the cutter are tried together for fit.

Stock removal in the keyslot is accomplished by the climb-milling method. Provision is made in the body for a generous flow of coolant so that clips are flushed away before they can get into the worm gearing and bearings. The shaft is drilled in line with its center for oil lubrication of bearings E and D. Auxiliary holes lead to the bearings. These bearings have grooves on their inside-diameter surfaces to distribute oil. The shaft oil hole leads to a radial filler hole on the shank of the driving shaft between the top bearing shoulder and the blend area of the Morse taper shank.

Floating Drive for Laps

Holes in hardened work are often finished to a precise and smooth surface with adjustable laps. It is difficult, however, to do a good job if the lap is not in accurate alignment with the hole. In practice, the average drill chuck usually does not run true enough to permit proper lapping. This problem can be solved by equipping a set of laps with a simple, free-floating drive, Fig. 5, which allows the lap to align accurately regardless of the errors in the chuck or the shank of the lap.

A heavy-duty diemaker's compression spring, Fig. 6, serves as the flexible drive link between the lap and a stud gripped in the chuck. Such a spring is strong enough to drive the tool, yet is flexible enough to permit the free-floating action which is necessary for alignment.

The spring is held on both the drive stud and the lap shank by a force fit. Both shanks are turned to about 0.001 inch over the inside diameter of the spring and the three parts are pressed together. Usually this degree of force fit is sufficient to prevent rotational slippage. If this is not the case and slippage between the various components should occur,

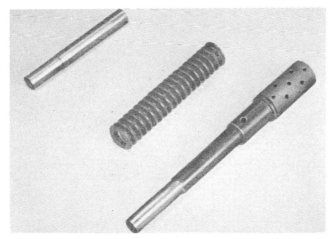

Fig. 6.　Drive stud (top), spring (center), and lap (bottom) are assembled to form the floating lap drive shown in Fig. 5.

each end of the spring can be tack-welded to the shanks of the drive stud and the lap with an arc welder.

Attachment that Adapts a Drilling Machine for Lapping

The mechanism illustrated in Fig. 7 was designed to impart an automatic reciprocating movement to the spindle of a drilling machine, so that the machine could be used for lapping operations. This attachment was originally applied to an old type of drilling machine, but with a few alterations it can be adapted to suit any type.

A worm, keyed to the revolving spindle, drives a worm-wheel on the shaft of a disc crank. Pins in the disc crank slide in guides that are firmly attached to the machine body. As a result, the pins can only move horizontally. This horizontal movement forces the disc crank slide and the gear into a vertical reciprocating movement. The slide, in turn, transmits this movement to the drill spindle, so that it moves up and down. The friction is taken up by ball bearings on each side of the worm. The slide is kept from revolving with the spindle by a guide bar, which is secured to the baseplate.

The ratio between the worm and worm-wheel is 1 to 13 in this case, resulting in 6 1/2 revolutions per stroke. The length of the stroke is about 5 inches. Provision for adjusting the length of stroke can be incorporated in the design if desired. To apply this attachment to a drill

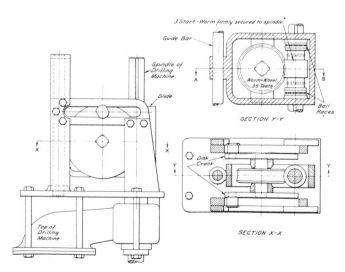

Fig. 7.   Drill press attachment employed to impart a vertical reciprocating movement to the spindle for performing lapping operations.

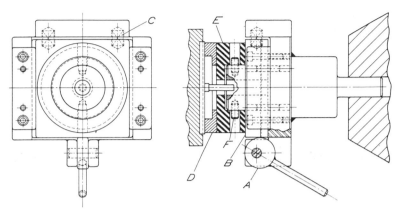

Fig. 8.   The floating action of rubber disc *E* permits the end of the work to be lapped despite the presence of a protrusion.

press, the pinion meshing with the rack of the spindle has to be removed, after which the whole structure is securely fastened to the machine body.

Surface with Protrusion Lapped on Hand Screw Machine

Lapping the end of a small cylinder posed a problem because of the presence of a protruding shaft in the center of the surface to be lapped. To do the job, the work was chucked in a small hand screw machine, and the device illustrated in Fig. 8 was set up in one of the stations of the hexagon turret. The view is cross sectional, looking down from above the setup.

Cam *A* is actuated by the operator to advance slide *B* against the pressure of springs *C*, which retract the slide when the cam is reversed. Cast-iron lap *D* is bonded to rubber disc *E*, which in turn is retained loosely by set-screws *F*. In use, the tool is brought up to the work with light pressure, and the cam is rotated back and forth until the proper finish is obtained.

Tooling for Precise Shaving of Stampings

Air-operated equipment that accurately shaves the edge of a stamping along a circular arc is illustrated in Fig. 9. The operation is performed by rotating the workpiece on a horizontal turntable past a series of fixed cutters. Although designed for a particular part, this arrangement may be adapted for similar applications.

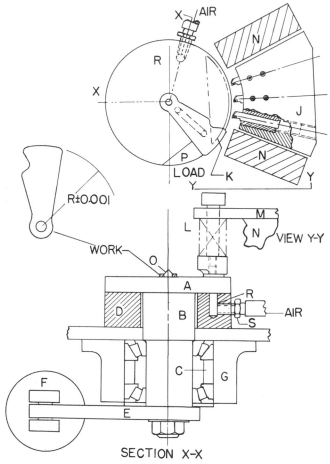

Fig. 9.   Device with an air-operated turntable for shaving the edges of stampings to obtain an accurate radius.

Built into the top of a cast-iron table, the device consists basically of a faceplate A mounted on a spindle B that rotates in precision bearings C. Under the faceplate and secured to the top of the table is a hardened disc D. The disc is in contact with the bottom of the faceplate.

The lower, threaded end of the spindle is keyed to an operating lever E. Retained by a nut and washer, this arm is rotated through an arc of about 130 degrees by means of an air-operated cylinder F which is clevis-mounted to the bottom of the table. The faceplate A, spindle B, and disc D are all precision ground from hardened tool steel. Suitable provision is made for lubrication between the contacting surfaces of members *A* and *D*. The bearings are mounted in a housing *G* which extends below the table top.

Tool-holder *J* carries three carbide-tipped tool bits. These tools have 3/8-inch-square shanks and are clamped in square-to-round adapter sleeves brazed into holes drilled in the tool-holder. Two set-screws locate and secure each of the tools. The bits are square-ended but have a top relief with the high side to the right.

Each tool cuts with a shearing action — the side thrust forcing the work against the faceplate rather than lifting it. The first tool is set to take a heavy roughing cut. The other two remove about 0.002 inch each to obtain a fine finish. An 0.010-inch stock allowance should be provided for trimming.

A spring-mounted pad *K*, machined in the form of a circular segment, wipes against the part to hold it in place during the cutting stroke. The pad is made of a hard copper-bronze for wear resistance and is beveled at the approach end. The thrust from the die springs *L* is taken by a bar *M*. This bar is bent into a U-shape, and mounted on two blocks *N*. Stripper bolts retain the pad.

Air ejection of the stamping at the end of the work stroke is accomplished by having two holes in the faceplate line up with an air passage in disc *D*. As the air cylinder piston pauses before snapping back to the load position, the workpiece is blown clear of the faceplate. A few simple sheet-metal guides direct the trimmed parts into tote pans.

In loading the device, the workpiece is positioned by a pin *O* in the center of the faceplate. A block *P* is also secured to the faceplate to serve as a drive for the part. The hold-down pad forces the stamping against the plate before it makes contact with the cutting tools.

With the part in place, the operating cycle is started by depressing a key to open the air valve. The air enters the cylinder and forces the piston-rod out to turn the faceplate in a counterclockwise direction. As the part rotates it slips under the beveled edge of the hold-down pad *K* and is trimmed by the three cutting tools. At the unload station the part is automatically ejected by air supplied to passage *R* through a hose connected by means of fitting *S*. The air cylinder then resets the device for the next stroke.

Simple Wire-straightening Device

A lathe attachment designed to straighten coiled wire or lengths of wire that have been cut to length is shown in Fig. 10. One shank end of the device is mounted in the lathe chuck and the other shank end is

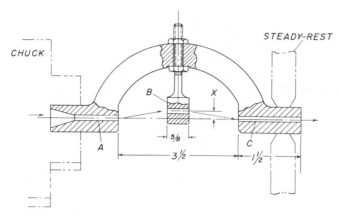

Fig. 10. Lathe attachment used for straightening wire from coils as well as cut-off lengths of wire.

supported in a steadyrest. The wire is threaded through the hollow head-stock spindle and then through holes A, B, and C of the attachment.

By running the lathe at a minimum speed of 1400 rpm and pulling on the wire with a pair of pliers, clean straight wire will be drawn from hole C. About 1500 feet of wire can be straightened per hour. Distance X, the amount that hole B is offset from the center line of holes A and C, must be adjusted for differences in the hardness and diameter of wire. The attachment has been used for wire up to 3/16 inch diameter but larger diameters can be handled.

Hand-operated Spring Winder Simple to Make

Coil springs can be wound manually at the bench with the easily constructed spring-winding arrangement illustrated in Fig. 11. Within its capacity, this device can be used to accurately produce compression and tension springs for short-run production.

Basically, the spring winder consists of a mandrel A mounted in a bushing B that is threaded into a support plate C. A tensioning screw D has a through-hole adjacent to the head to pass the spring wire E, and coil compression spring F is mounted on the body of this screw for the purpose of applying the necessary pressure. Spring spacer G is circular and wedge-shaped, as shown, so that it can be rotated to obtain springs of any pitch within its range. The spring spacer is held in place by a cap-screw which can be positioned anywhere along a slot in the support plate. This arrangement permits the winding of springs of various diameters.

In operation, the wire is inserted through the hole in the tensioning screw and then placed in the slot at the end of the mandrel. The spring spacer G is brought up against the mandrel, rotated to the position that will give the desired pitch to the spring, and secured in place with the cap-screw. After adjusting the pressure on the tensioning screw by means of the nut, the mandrel is cranked by hand in the desired direction to wind the spring. For each diameter spring, a different mandrel and bushing are required. The side of the mandrel at which the spring spacer is set is determined by the direction the spring is wound.

Double-Coil Winding Fixture

In a recent design of an insecticidal fog application machine, maximum efficiency required that the liquid be directed through a cooling coil of copper tubing on the suction side of the pump. Limited space precluded the use of a single coil of sufficient length, so the tubing was wound in a continuous double coil — an inner coil and an outer coil. A lathe was adapted with the fixture shown in Fig. 12 for producing the coil.

This fixture consists of a cast-iron faceplate A, having an integral cored hub B, a bar C, and a sleeve D. The bar, which is the same diameter as the required inside diameter of the inner coil, is permanently fixed to the faceplate; and the sleeve, having the same outside diameter as the inside diameter of the outer coil, fits on the machined core of the faceplate hub, and is removable.

Winding of the inner coil is completed first, one end of the tubing being held in a hole E drilled in the faceplate. As the lathe spindle rotates slowly, a hook F projecting from a point on the bar near the hub catches

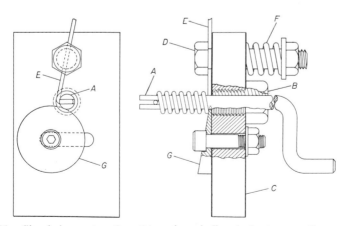

Fig. 11. Simple in construction, this spring-winding device is manually operated.

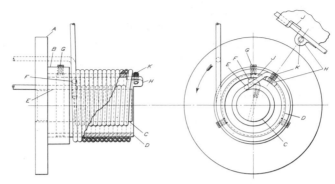

Fig. 12. Fixture mounted on a lathe for winding a continuous double coil of copper tubing.

the tubing. A suitable supporting member in the toolpost guides the tubing onto the bar as the carriage moves to the right. This movement is controlled by the lead-screw of the lathe, which has been geared for 5 1/2 threads per inch, thus producing a lead of 0.181 inch. Since this lead corresponds quite closely to the outside diameter of the tubing — 0.187 inch — a tightly wound coil is assured.

When the tubing has been completely wound to the right-hand end of the bar, the lathe is momentarily stopped, and sleeve D is slipped over the still unwound length of tubing and into position over the inner coil. Set-screws G secure the sleeve to the hub. The operator then directs the tubing over the lip H on the end of the sleeve and the tubing winds around the outside diameter of the sleeve as the lathe spindle rotates and the carriage moves to the left. The enlarged view of the lip shows how the tubing runs out through a slot J and back against a threaded end-stop K.

After the outer coil has been completed, the lathe is stopped, the set-screws G are loosened, and the coils and sleeve are removed as a unit. The coils can then be slipped off the faceplate end of the sleeve. Since the lathe spindle rotates in the same direction for winding both coils, the inner coil has a left-hand helix and the outer coil, a right-hand helix.

High-Speed Riveting Attachment for Drill Press

A high-speed riveting device has been designed for use with a conventional drill press. The drive shank of the unit is tapered to fit the drill press spindle from which it derives its power. Changes in the rotational speed of the spindle will proportionally affect the rapidity with which the peening strokes occur.

The components of the tool, shown in Fig. 13, fit within a tubular steel body A that is bored out to five different diameters. Located within one of the step bores is a grooved spindle B. The largest portion of the spindle is 1/16 inch smaller in diameter than its corresponding body bore. A needle bearing C maintains concentricity.

The upper end of the body bore is threaded to receive cap D which is screwed snugly against the bore shoulder. A hole is bored through the cap enabling it to slip over the spindle shank. For tightening purposes, a series of blind holes to accommodate a spanner wrench is also provided. A ball-thrust bearing is housed within a counterbore in the lower face of the cap.

Situated immediately beneath the spindle is sleeve E, cross-pinned to the small end of rod F. Rotation of these two members within the body is prevented by key G which restricts movement to a reciprocating motion. Approximately 3/8 inch is the limit of the sliding movement in this case. Coil spring H maintains the two members in their normal raised position as shown.

An adjusting-screw J, a close sliding fit over the lower end of rod F, is threaded into a recess in the conical end of the body. A dog point set-screw K passes through the adjusting-screw flange radially and extends into an annular groove cut around the circumference of the rod. The width of the groove is equal to the sum of the set-screw dog diameter and the maximum amount of sliding movement possessed by the sleeve and rod. Circular lock-nut L is threaded on the shank of the adjusting-screw to provide a means of securing it to any desired setting. The peripheries of both the screw and nut are knurled to facilitate gripping. By means of this set-up, the length of riveting stroke imparted to rod F may be altered as needed to suit various requirements.

Formed integrally on the lower end of spindle B is node M, 3/8 inch high and 1/2 inch thick. The leading side of the node (left-hand side) is inclined steeply, while the trailing side is inclined at a lesser angle. The crest is rounded, polished, and heat-treated to withstand the impact to which it will be subjected.

A similar integral node N is provided on the upper end of the sleeve. Although both sides of this protuberance are inclined at an angle of 45 degrees, its height and width are the same as for node M. Radial locations from the center of the spindle must be identical for both. The lower end of rod F is bored to receive the round shank of rivet-set O. A cross-pin holds it in place.

Before using this unit, cap D should be removed and an ample quantity of grease stuffed into the body around the revolving spindle and the

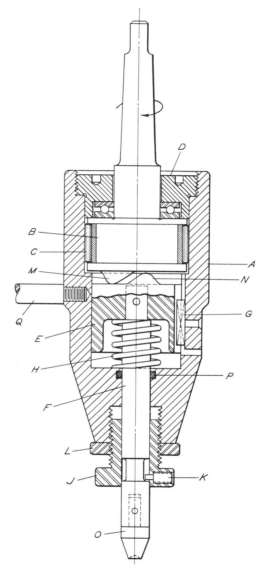

Fig. 13.   Riveting attachment for drill press provides one blow for each revolution of the spindle.

reciprocating members. Grease seal *P* prevents the escape of lubricant through the nose of the device.

In operation, the tapered shank of the tool is inserted into the drill press spindle just as would be done with any tapered-shank twist drill.

Because body A must remain stationary, an arm Q is screwed into a blind tapped hole in the body wall and braced against the machine frame. The machine table is then adjusted to the proper height.

When the drill press is started in the normal direction of rotation, spindle B will be driven in the direction indicated by the arrow. Once during each revolution of the spindle, node M will strike node N. As a result of this, sleeve E and rod F will be driven sharply downward. The sleeve and rod will move 3/8 inch when the tool is set for its maximum stroke. After the upper node has passed this point, the return spring will once again raise the sleeve and rod to their original positions. Optimum operating conditions may be realized within the speed range of 200 to 250 rpm.

Self-locking Driver for Parts with Socket Head

An ingenious self-locking device which utilizes three balls for driving long cylindrical parts having a hexagonal socket in one end is shown in Fig. 14. The device is intended for use in a lathe equipped with an air-actuated tailstock.

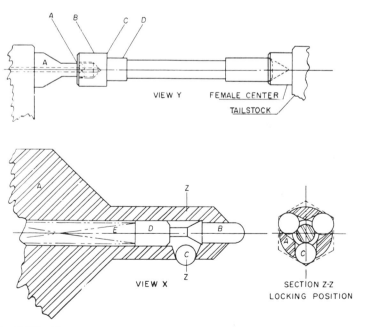

Fig. 14.   Self-locking device designed for driving parts having a hexagonal socket in one end.

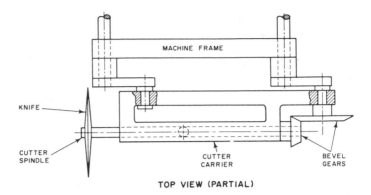

TOP VIEW (PARTIAL)

SIDE VIEW

Fig. 15. High-speed cutoff device for strip material.

View Y illustrates the self-locking driver being applied for holding a part that is machined on surfaces *A*, *B*, *C*, and *D*. In loading the work, the socket end of the work is advanced over the reduced section of adapter *A* until the bottom of the socket hole has contacted plunger *B*, view X. The plunger is pushed back into adapter *A* until its tapered section has forced balls *C* radially outward to lock them against the sides of the hexagonal socket.

When the tailstock spindle is retracted, coil spring *E* returns plug *D* and plunger *B* forward, releasing balls *C* so that the socket end of the work can be removed from adapter *A*.

With this method of driving, parts with hexagonal sockets are positively held, concentric with the centers of the headstock and tailstock.

High-Speed Cutoff Device

Continuously moving strip material can be cut without stopping the strip, thus speeding production. Conventionally, this type of cutting is done while moving the tool at the same velocity as the material and then withdrawing the cutter and returning it to the starting position and repeating the operation.

The principle is illustrated by the device shown in Fig. 15. The material to be cut moves at constant velocity. A circular knife on a shaft is supported by a cutter carrier which swings in a circular path on a pair of equal-length cranks. The cranks are driven clockwise at the same angular velocity by one or both of the connected crankshafts.

When the right-hand crank is rotated, a bevel gear fixed to the crank drives a second gear fixed to the cutter spindle and revolves the knife. The cutter carrier has a nonrotating shaft which imparts a reciprocating motion through a sleeve bushing to a slide supporting the material while it is being cut.

Since, in this arrangement, the cutter and work move in the same direction at the same speed, it is not possible to change the length of the cut-off piece without reproportioning the parts of the device.

Tool for Assembling O-Rings in Deep Bores

Although in most applications O-rings are assembled to the shaft, there are occasions when it is necessary to put them in a deep or inaccessible bore. For example, wet-sleeve diesel cylinder liners are usually so thin-walled that it is impossible to groove them for O-rings. Therefore, the O-rings are put in the cylinder block. The tool here described was developed and used in production for a job of this type.

The tool (Fig. 16) consists of a shaft A on which is pressed an upper pilot B and a lower pilot C. Two grooves, machined in member C, are spaced the same distance apart as the O-ring grooves in the block. The grooves in pilot C are made slightly less in width than the diameter of the O-rings in order to hold them in the pilot until installation. Four semicircular shoes D are located at the bottom of the grooves. These shoes, which are a slip fit in the grooves, are restrained from falling out by stop-pins E.

Bearing on the center of each shoe is a plunger F that is a close fit in a reamed hole in pilot C. Center shaft A is drilled through and tapped for an air connection, and cross-holes are drilled to allow the air to enter at the underside of plungers F. Also, a notch is cut through pilot C, as shown in Fig. 17.

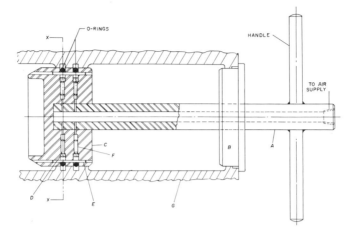

Fig. 16.   Air-operated tool for installing O-rings in a deep bore.

To use the device, O-rings are placed in the two grooves in pilot *C* (Fig. 17) and the slack in each ring is tucked into the notch. The tool is then inserted in the bore of the workpiece *G*. Pilots *B* and *C* center the tool, and the shoulder on member *B* stops against the top edge of the bore, thereby insuring that the O-rings are exactly opposite the grooves

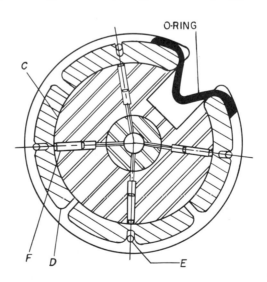

SECTION X-X

Fig. 17.   Cross section of tool seen in Fig. 16 showing how O-ring is mounted on pilot *C*.

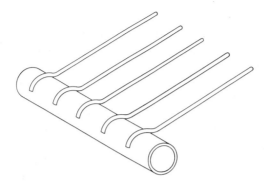

Fig. 18.   Welded assembly that is produced automatically on a resistance welder.

in the part. At this point the air is turned on and the pressure applied through the center shaft *A*, forcing plungers *F* against shoes *D*. These shoes in turn push the O-rings into the grooves in the part. The tool is rotated about one-quarter turn to "iron" the O-rings into the grooves before the air pressure is shut off and the tool withdrawn. In the setup used, the air was controlled by a foot-operated valve, thus freeing the operator's hands.

Resistance Welder Automated for Multiple Operations

An arrangement for the automatic positioning, cutting, forming, and welding of the parts of a simple assembly is here illustrated. The work piece, Fig. 18, is a 6-inch length of 3/4-inch steel tubing *A*, Fig. 19, having five pieces of 3/16-inch diameter wire *B* formed and welded to the periphery. The device is used in combination with a standard 100-kva resistance welder.

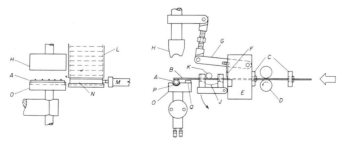

Fig. 19.   Setup of special tooling for automation of a resistance welder to produce the assembly seen in Fig. 18. In addition to the welding operation, the wires are cut to length and formed to fit the tubing.

Tooling consists of three basic sections: a stock feed mechanism for the five wires, with cutoff blades and a roller to hold the wires in position after the cutoff operation; a hopper feed for the lengths of tubing; and the welding station. The wire feed arrangement is mounted on the welder but is supported independently, as is the hopper feed for the tubing. Five horizontal power-feed reels (not shown) each hold a coil of wire. From these reels, the five wires are led through power straighteners (not shown) and two bushing plates C, each having five guide bushings. Between these plates is a pair of rubber-covered feed rolls D which are synchronized by a roller chain and sprockets. The rolls are held together under spring pressure and the lower of the two rolls is driven by an air cylinder, through a rack and pinion and an over-running clutch.

After passing through the guide bushings in the second of the bushing plates, the wires enter a cutoff die E. The lower die blade is fixed and the upper blade F is operated by a long lever G at a mechanical advantage of 5 to 1. One end of this lever is actuated by the pressure cylinder for the upper electrode H of the welder. By arranging steps and the shear angle on the cutoff blade, the five 3/16-inch wires are cut off smoothly without a sudden breakthrough to adversely affect the welder.

From the cutoff blades, the five wires lead through two bronze guides J and under a rubber pressure roll K positioned between the guides. Both guides are cast with wide openings on the wire-entrance side to automatically align the wire as it feeds in from the cutoff die. The rubber pressure roller holds the wires in the guide slots, during and after the cutoff operation, until the weld is completed, and also prevents the wires from shifting lengthwise. This roller is covered with soft rubber and is mounted on the frame of the cutoff die. The two forward guides J are secured to a hinged plate that swings down to allow ejection of the finished part.

The hopper feed device L for the lengths of tubing is mounted on the front of the machine, and an air cylinder M is secured to one end of the hopper. When the air cylinder is triggered, a cross-bar on the end of the piston-rod pushes the bottom tube N in the hopper forward into the lower work-holding electrode O. This automatically forces the previously completed part out of the machine. When the piston of the hopper cylinder retracts, another tube drops into position, ready for the next feed stroke.

The welding station consists of a simple, elongated upper electrode that both forms and welds the wire, and the lower electrode which holds the tube in a semicircular nest formed of flat, braided copper cable P brazed to the lower electrode. Under welding pressure, the cable crushes down to form a perfect contact with the tubing. This eliminates both deformation of the tubing and poor welds from improper contact between the tube and the electrode. One side of the lower electrode holds a carbon block Q that assists the upper electrode as it forms the wires around the tubing. In addition, the carbon block supports the wires after the front guides have been automatically lowered for ejection of part.

To start operation, the five wire reels are loaded, and the wires are run through the straighteners and the bushings in the first bushing plate, up to the rubber feed rolls. With the welding circuit open, the machine is intermittently operated until the feed rolls have gripped all five wires. They are then threaded through the second bushing plate and the cutoff die. The machine is slowly operated through the rest of the cycle, leaving the ends of the five wires trimmed off and even. Then, after setting the machine at rest, the tube hopper is filled, the welding circuit is energized, and the cooling water is circulated.

When the control is set on automatic and the start button is depressed, the air cylinder on the hopper feed operates, causing a tube to be pushed into the lower electrode. At the same time, the air cylinder drives the wire feed rolls, advancing the wires through the front guides and over the tube. The air cylinder pistons of both the hopper feed and the feed rolls retract, and the welding head starts down, activating the die blade which cuts off the five wires. Held in place by the pressure roll between the guides, the wires are contacted by the upper electrode which forms them around the tube. Then the welding cycle begins. As soon as the part is welded, the front wire guides lower away from the wire and the upper electrode rises. The hopper feed cylinder pushes in a new tube, forcing out the completed assembly as the feed rolls start feeding more wire forward. As soon as the hopper feed piston reaches the end of its outward stroke, an air cylinder pops the front wire guides back into place and the cycle repeats.

Improved Hubbing Technique for Plastic Molds

Hubbing of cavities in plastic molds, or die-casting dies, is often facilitated by employing a guide between the hub and the blank. This practice insures correct alignment between the tool and the work. In the case of round workpieces, a strong bushing is prepared and both the hub and the blank are turned to a sliding fit with the hole of the bushing. The design shown at A (Fig. 20) has a serious drawback, as the work-blank always expands considerably during the hubbing process. This makes it difficult to get the workpiece out of the guide.

A simple solution for this problem is shown at B. The bushing is of two-piece construction and inserted into a larger conventional bushing or

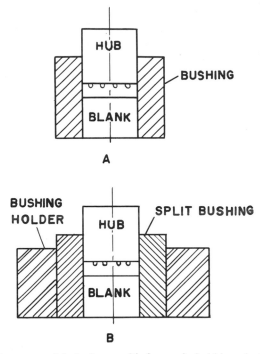

Fig. 20.   Two ways of designing a guide for use in hubbing plastic molds.

holder. With this design, the split bushing can be easily extracted from the holder and the blank freed immediately regardless of the amount of blank expansion.

Circular Gasket Cutter

A tool for cutting circular gaskets or washers makes use of two throw-away surgical knife blades. This circle cutter is excellent for small-lot production or even single-piece jobs. It is especially valuable in experimental or model shops. The blades have the extremely sharp edges needed in cutting gasket materials, thin rubber, and Teflon. Blade cost is low when compared with the time required to sharpen other types of cutting tools.

As shown in Fig. 21, the cutter body consists of a cylindrical shank A, which can be gripped in a drill chuck, and a head B milled along two opposite sides to a keystone shape. Each milled surface serves as a base for a cutter block C.

The blocks are tongued with their respective surfaces and secured by cap-screws D. Slots E in the blocks, through which the cap-screws run, provide the means of setting the blades to the gasket size desired.

At one end, each block is milled a close fit with the surgical knife blade F. Set-screws G clamp the blade to the block. One of the blades appears in the upper right view of Fig. 21. Only the bottom part of the blade is used, the top part being cut off or chopped off and discarded.

The blades in the tool illustrated are the same distance from the center line, and will cut a solid disc of some specified diameter. To cut a ring or a washer, the cutter blocks are adjusted radially so that one blade will cut the outside diameter and the other, the inside diameter. The toed-in angle of the blocks makes it possible to cut very small circles. A small

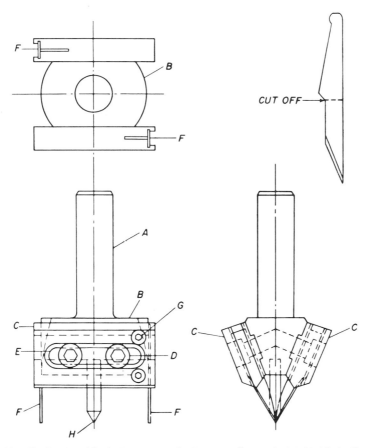

Fig. 21.   Each cutter block C supports the bottom of a surgical knife blade F and can be adjusted radially in head B for the gasket circle required.

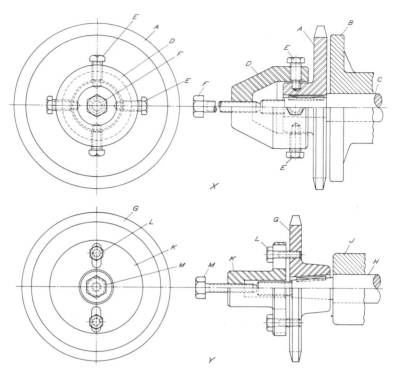

Fig. 22.   Two types of wheel-pullers designed for operation in close quarters.

conical center *H* in the body keeps the gasket material from moving while being cut.

Wheel-Pullers Function in Restricted Locations

Effective application of conventional claw type wheel-pullers may be hindered by the proximity of other components which prevent adequate gripping of the claws or by relatively frail proportions of the wheel flange to be gripped. Two types of wheel-pullers designed to function satisfactorily in tight quarters, without buckling or otherwise damaging the wheel flange, are illustrated in Fig. 22.

A typical sprocket wheel *A*, with its flange situated close to bearing bracket *B*, is keyed to shaft *C*, as shown at X. Body *D* of the extractor, which is machined from a steel bar, passes freely over the sprocket-wheel hub. Four through holes are drilled and tapped equidistantly around the right-hand end of the body. Fitted into each of these holes is a hardened steel hexagon screw *E*. The diameter at the tip of each screw shank is reduced to below the thread roots, forming a cylindrical dog. These dogs pass freely into shallow blind holes that are drilled into sprocket hub for extraction purposes only.

The end wall of the body is drilled and tapped to receive hardened steel pressure-screw F. The conical tip of this screw bears against the end of shaft C, With the dog ends of screws E engaging the holes in the hub as shown, pressure-screw F is rotated to exert pressure against the end of the shaft. This forces body D to the left, taking sprocket wheel A with it.

With similar types of sprocket wheels, mounted in this way, it is a simple matter to spot-drill the four blind holes around the boss. By removing the four screws E, body D may be used as a template.

At Y can be seen an alternate design of wheel-puller intended for use in cases where a sprocket wheel or gear is mounted with its outside flange face accessible. In this example, sprocket wheel G is keyed to the tapered end of shaft H, which is supported by bearing bracket J. The flanged body K of the extractor is bored out along part of its length so as to slip over the smaller diameter at the end of the shaft.

Two slots, 180 degrees apart, are machined through the flange of body K to receive hexagon screws L. These screws are threaded into mounting holes that have been drilled and tapped through the flange of the sprocket wheel. The end wall of the extractor body is drilled and tapped to receive hardened steel pressure-screw M. The conical tip of the pressure-screw bears against the end of shaft H.

With the two screws L threaded into place, and pressure-screw M situated as shown, the wheel-puller is ready for use. By tightening the pressure-screw with a wrench, both the extractor body and the sprocket wheel will be forced to the left.

Device for Levelling Bored Parts

When using a bored hole as a datum, the simple levelling arrangement illustrated in Figs. 23 and 24 may prove effective in the setting up of

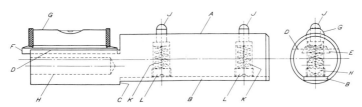

Fig. 23. Details of plug mounting for spirit level that permits easy setup of parts from bored holes. Spring-loaded plungers J keep device aligned with hole.

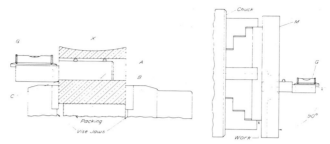

Fig. 24. Part held in machine vise is tested for levelness. Special bar *M* adapts the device for setting up thin, flat work in a lathe chuck.

rough or irregular workpieces for machining. The device employs a spirit level and is applied either directly to the part or indirectly through the use of an auxiliary mounting.

The body of the device is illustrated at *A* in Fig. 23. It is a cylindrical rod of casehardened steel that has been ground to have an outside diameter of about 1/8 inch less than the smallest hole in the work being used as a datum. The length of the rod is approximately one and one-half that of the hole. A rectangular slot *B* is milled along and parallel to one side of the rod, extending about two-thirds the length of the part. A square step *C* is formed at the end of the slot.

Flat *D* is machined into the other end of the rod and diametrically opposite the slot. This flat extends almost to step *C*, which is located on the underside. A dovetail groove *E* is milled centrally along the width of the flat. The two long sides of the base *F* of a spirit level *G* are filed at an angle to be a light press-fit into the dovetail groove. To reduce the weight of the device, a large-diameter, blind hole *H* is drilled into this end of the rod.

Two hardened steel plungers *J* slide freely within holes bored diametrically through the right end of the rod. These plungers are situated a considerable distance apart and are perpendicular to the base of the slot. The end of each plunger projects about 1/4 inch out of the rod and is rounded hemispherically. A light coil spring *K* bears against the head of each plunger to force it outward. Each spring, in turn, acts against a threaded plug *L* that is screwed into the lower end of the guide hole.

The way the levelling device is employed to align a workpiece *X* that is gripped in a vise is shown at the left in Fig. 24. The part of the rod extending beyond the spirit level is simply inserted into the hole in the casting before the latter is gripped in the vise of the machine. Rod *A* is placed so that slot *B* is at the lower side of the hole and step *C* abuts the front end face as shown. The edges of the slot bear evenly against the side of the hole to prevent the rod from turning easily. When the device is inserted into the casting, the spring pressure on the depressed plungers serves to hold the rod firmly in alignment with the center axis of the hole.

The casting is then clamped in the vise on the unmachined surfaces and adjusted until the spirit level indicates a truly horizontal setting. Care should be taken to avoid closing the vise jaws on a very uneven surface or on high spots.

Another application of the levelling device is depicted at the right in Fig. 24. In this case, it is employed to check the setup of a large-diameter, rough, ring casting held in a lathe chuck for boring and facing. It will be noted that the illustrated casting is narrower than the widths of steps on the chuck jaws. The workpiece, therefore, cannot be positioned against the jaws for facing.

Ordinarily, four blocks of identical height would be inserted behind the workpiece — one at each jaw. Blocks of this kind are rarely at hand when required and are difficult to manipulate and retain in position behind the workpiece until a tight grip is obtained on the casting.

One adaptation of this device that avoids these setup difficulties is comprised of the complete levelling gage and a hardened and ground rectangular steel bar M. The sides of the bar are ground flat and parallel to each other. Being long enough to extend across the full diameter of the largest ring casting to be machined, the bar is bored somewhere near the center of the length to receive the cylindrical rod A. The hole must be machined perpendicular to the bar sides.

When the attachment is inserted, the foremost plunger J holds the levelling device firmly in position. The width of bar M need only be about one and one-half times the outside diameter of the rod. To reduce the weight of the bar a slot is machined into the front side. This slot extends almost the full length of the member, but is terminated close to each side of the hole, as indicated by the broken lines.

In use, the casting is first lightly gripped in the four-jaw chuck and allowed to project a distance sufficient for machining. The bar with the levelling device mounted is placed, preferably in a vertical position, against the face of the workpiece. Then the casting is tapped at top or bottom until the spirit level shows a horizontal setting. The chuck and workpiece are revolved by hand about a quarter revolution to make a second similar check, with bar M again held in the vertical position. When the level records the horizontal settings at both positions, the casting will be gripped true.

Socket-Wrench Manipulator

Wherever cap-screws have to be tightened in areas where no sweep of a handle is permissible, a special tool has proved helpful. The device offers straight-line wrenching as well as torque at any angle through 90 degrees to the axis of the cap-screw.

The operator holds knob *A* (see Fig. 25) with his right hand and housing *K* with his left hand to place the proper-sized wrench socket over the head of the cap-screw. The differential action of the gear train

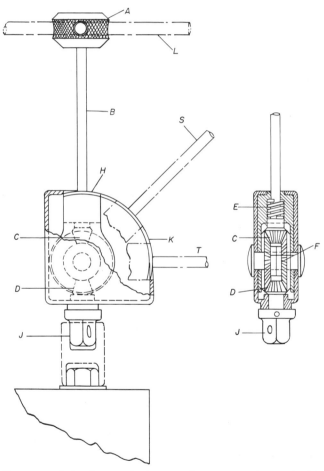

Fig. 25. Cap-screws in hard-to-reach places can be run up and securely tightened with this novel differential device. The view on the left shows how the handle can be shifted out of line as far as 90 degrees and deliver full torque.

C to F to D is such that shaft B must be turned counterclockwise (left-handed) to "make up" a screw, and clockwise (right-handed) to back out a screw.

When it is desired to run up a screw off the perpendicular — for example, to position S or T — the adjustment is made by pulling on shaft B. Pinion C disengages from the mitre gears F so that the desired angle can be set. By releasing shaft B, compression spring E drives the pinion back into mesh. Pinion D is fixed on shaft J, always in mesh with mitre gears F. The gear assembly is held in place within the housing by yoke H so that the angle shift is accomplished with ease and speed.

Knob A is drilled through with holes at right angles so that windlass lever bar L can be run through the holes to achieve more leverage.

CHAPTER 16

Tooling Component Design—General

Simplified Chucking-indexing Fixture

For speed, simplicity, and economy it is possible to build a chucking and indexing fixture which holds tubular parts by the inside diameter, locating on the outside diameter. Inception of this fixture was to mill eight keyways radially across a flange face (see Fig. 3, lower left).

The fixture can be used in any position, and is so constructed that the blank is held using special collet C, Fig. 1. Work-rest A eliminates the need for gaging depth dimensions of the parts to be machined. When loading the chuck, the piece-part is pulled toward the rest, making the work seat securely. To hold other circular shapes, such as the tubing in Fig. 2, the part seats in the nest. Similarly, other parts can be chucked on millers and drill presses by making special nests, gages, and collets to suit the contours of the work. The indexing ratchet (top of Fig. 1) is linked to sprocket I by roller chain, Fig. 2. The chain is not shown.

This fixture can be built as a twin by connecting sprocket I (Fig. 1), by a second roller chain to a mate, as suggested by the broken outlines near the top of Fig. 2.

As illustrated in Fig. 1, tapered roller bearings F and thrust bearing P assure reliable repetitive indexing whether the collet is open or closed. Sprocket I and indexing disc J are keyed to main shaft K. Collet extension shaft L is held by a screw to collet expander B. Thrust bearing P is bolted to its housing M with bolt N (made to suit the thrust bearing's inside diameter) and lock-nut O. To open the chuck a lever linkage connects with a fork which rides in the slot of thrust housing M.

A hole is drilled at point R in shaft K for entry of grease to extension shaft L. Heavy washer E acts as a shield to prevent entry of chips to the bearings. Two lock-nuts at D are arranged to adjust the loading of the bearings. The housing is made of tank plate or SAE 1020 steel.

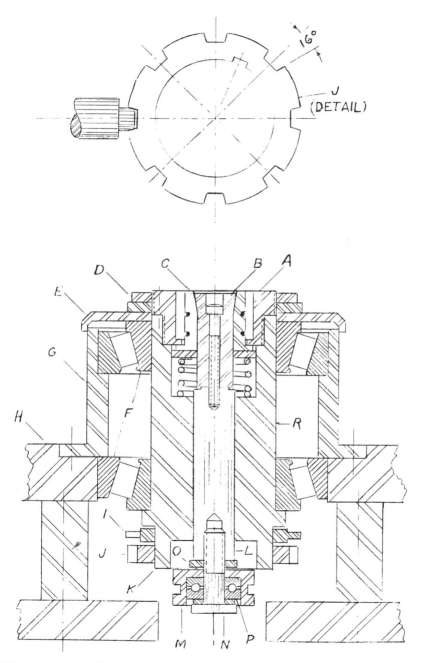

Fig. 1.   Extension shaft *L* rises by pressure of lever fingers in the slot of housing *M* to open three-jaw collet *C*.   Stop plunger for index is shown in the top view.

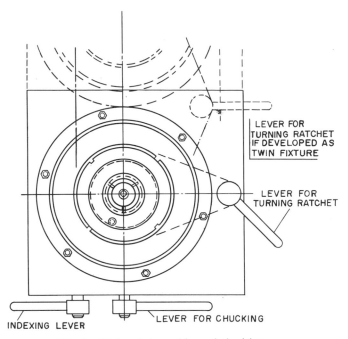

LEVER FOR
TURNING RATCHET
IF DEVELOPED AS
TWIN FIXTURE

LEVER FOR
TURNING RATCHET

INDEXING LEVER

LEVER FOR CHUCKING

Fig. 2.   Fixture is turned by a chain drive.

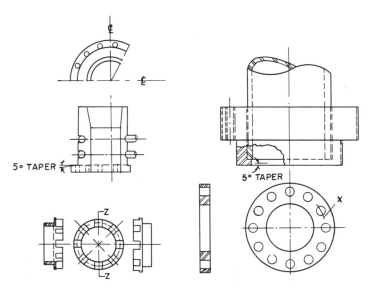

5° TAPER

5° TAPER

Fig. 3.   Three-jaw collet, at the left, has air jets for blowing chips away. Tubing is held in nest, as at the right. At the lower left is a typical workpiece, and at the right is a detail sketch of the collet support washer.

This fixture can be made with the equipment in any small toolroom, and results in a large saving in time and money.

Air Indexing Fixture

A dividing head type fixture combines the accuracy of indexing-pin engagement and the speed of air actuation. It was developed principally for work where weight and angular location are important factors. The entire operation is directed from a single control lever.

Design features are illustrated in Fig. 4. Index-pin lever *A* is shown in solid outline (position I) when the pin proper is engaged in one of the holes of index-plate *B*. When the lever is pivoted to the left as shown in broken outline (position II), the pin is withdrawn from the plate and the stem of the lever reverses two-way valve *C*, directing air to double-acting cylinder *D*. Rod *E*, connected to a piston in the cylinder, thus strokes to the right.

Carried by the rod is sliding key *F*, cut at a 45-degree angle at each end. When the rod is retracted to the left, a tension spring keeps the key normally raised, as shown in solid outline (position III) where it half enters one of the notches in wheel *G*. This wheel and index-plate *B* are secured to a common shaft and rotate as a unit. As the rod strokes to the right, the key causes the wheel to rotate and the notch, in turn, encloses the end of the key, forcing the key to fall in its slide.

When the leading side of the notch is perpendicular, the lower end of the key abuts inclined stop-block *H* (position IV), and rotation of the

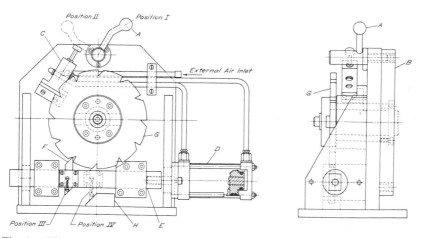

Fig. 4.   When lever *A* is pivoted from position I to II, key *F* moves transversely from position III to IV.

wheel is arrested. Lever A is then pivoted to position I and its index-pin engages the next hole in the plate. This movement of the lever again reverses valve C, retracting rod E to the left and letting key F spring up to half enter the next notch (position III). The indexing cycle is now complete, with the fixture locked for machining.

Wheel G can be designed with an odd or even number of equally spaced notches, according to the required number of work divisions. The wheel and the index-plate are related radially to each other so that when the key is in position IV, a hole in the index-plate is aligned with the index-pin. Actually, if the device is constructed precisely and stop-block H positioned correctly, repetitive indexing can be held to within 0.001 inch accuracy without engagement of the index-pin. With positive index-pin engagement, accuracy is even greater.

The design of the index-pin is illustrated in Fig. 5. Its movement is similar to the bolt action mechanism in a rifle. A spiral wringing motion of the lever in cam-slot J provides easy entry and withdrawal of the index-pin K from index-plate B. Since the assembly is spring-loaded, the pin when engaged is secured in place against slippage from machine vibrations. The design of the cam-slot permits the pin to be locked when

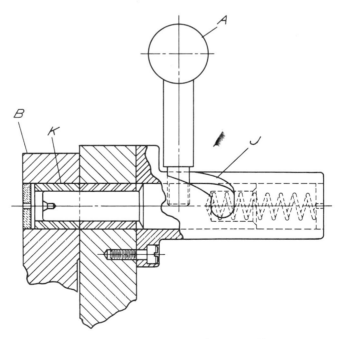

Fig. 5.   Like the bolt action of a rifle, pin K moves in or out of index plate B when lever A is pivoted.

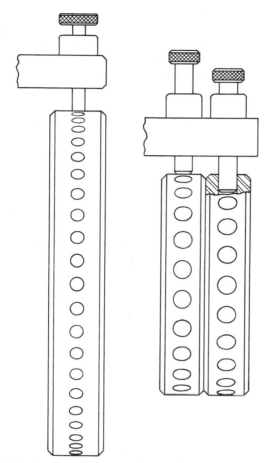

Fig. 6.   Diagram showing how two small index plates can be substituted for a cumbersome large plate.

withdrawn if it is desired to give a free movement to the indexing head.

Two Small Index-Plates Replace Conventional Large Plate

Often a pin-to-hole type index-plate is excessively large in diameter when used to provide a great number of divisions. Two smaller index-plates, each having half of the total number of holes in the big index-plate, can be substituted for the large plate. These plates are then accurately aligned, with the holes in one disc staggered with respect to the holes in the other disc. Figure 6 shows such a double-plate arrangement. The discs are fastened together with cap-screws and dowel-pins. In place of the original single index-pin two accurately aligned index-pins are used to engage the holes in the plates.

The usual spring type detent for the pins is not needed with this arrangement, since one pin will ride on the peripheral surface of its plate while the other pin is engaged. The additional space between the holes permits a larger pin diameter to be used; thus providing increased strength for the indexing mechanism.

Vise Jaws Serve as Adjustable Sine Bar Holding Fixture

A sine bar is the heart of a versatile, quickly adjustable fixture intended for the rapid setup of parts on which angular cuts are to be taken. The fixture, which is formed between the jaws of a machinist's vise, can be adjusted to conform to any angle by simply changing the height of the gage-blocks or the planer gage that supports the sine bar.

Under certain circumstances, this adjustable nest, which is shown in Fig. 7, also can be used to support parts in position for compound angle cuts. Such a case might be when only a single-tilting vise or a plain sine plate is available. The second, or compound, angle then can be set transversely in the vise by means of the setup shown.

View X is a cross-section through a vise, looking toward fixed jaw insert A. A pattern of holes has been provided in the jaw insert at a known distance Z from the horizontal inner surface of the vise.

A short length of steel rod is screwed to each end of sine bar B — rod C at the lower end and rod D at the upper end. Rod C will fit snugly in any of the holes in the jaw insert, thus securely anchoring the lower

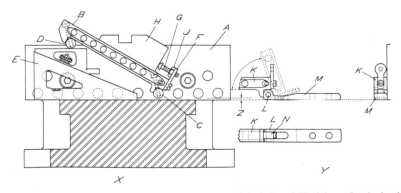

Fig. 7.   The lower end of sine bar B is supported in holes drilled into fixed vise jaw insert A, while the upper end rest on planer gage E. Easy adjustment of the sine bar and rapid loading and unloading of workpieces are features of this holding fixture.

end of the sine bar in a known position. Either gage-blocks or a planer gage E can be used to set the height of the upper end of sine bar B. It should be remembered that distance Z must be added to the normal height of the planer gage.

Stop-plate F is fastened to the lower end of the sine bar by a screw. Through the plate passes an adjustable stop-screw G which is used to position workpiece H at the desired height relative to the top of the vise jaws. Nut J will lock the screw at the proper setting. The height of the workpiece can also be altered by inserting rod C in a hole further up on the jaw insert. In this instance, a new dimension must be substituted for Z.

In the event that holes cannot be provided easily in the vise jaw insert, a mobile hinged device, view Y, can be used. Here, sine bar K pivots around a hinge pin L which is supported by a clevis type hinge arm M. The hinge pin passes through a hole bored in the center of short rod N that is attached to the sine bar, fitting between the fingers of the clevised end of the hinge arm.

The hinged sine-bar unit can be placed between the jaws of a vise. Height adjustment of the unit as a whole is made by placing narrow parallels beneath hinge arm M. The arm and the parallels can be held in place by a standard hold-down clamp and a filler block with a bolt coming up from the T-slot in the machine table. Two clearance holes are provided in the hinge in case it should be necessary to bolt it directly to another surface such as a T-slot block in the work-table.

When the sine bar is used in the hinged form and is set up on parallels, the gage-blocks or planer gage that is used to support the upper end can be placed on the parallels also. In this case, no allowance should be made for distance Z or any other distance, as the sine bar is in a normal position, contacting the surface of the same parallel as is the planer gage.

Fast-acting Retractable Locator

It often becomes necessary, on fixtures and dies, to provide one or more pins that can locate or hold a part while an operation is being performed, then may be pulled back to permit loading and unloading. Although there are numerous ways of doing this, many are slow-acting and not entirely positive.

Figure 8 shows a simple arrangement that overcomes most of these objections, yet is relatively inexpensive to construct. Locating pin A should be made of drill rod or tool steel, hardened and ground. Its front end is chamfered to facilitate entry into the workpiece, and the opposite end is tapped to receive stud B.

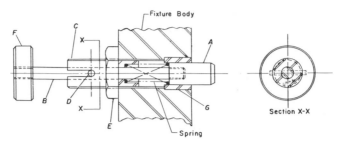

Fig. 8.   Pulling knob *F* retracts this locating pin. It will remain retracted if the knob is turned to bring dowel-pin *D* over the end of housing *C*. A spring is used for the return movement.

A round, externally threaded housing *C* is drilled through the center to accept stud *B*. The front end of the housing is counterbored to serve as a spring pocket, while the other end is slotted to receive a small dowel-pin *D* that is pressed into the stud. The entire unit can be assembled before inserting it into the fixture body. A lock-nut *E* secures the unit in place.

To load the fixture, the operator pulls back on knob *F* until dowel-pin *D* is out of the slot. A slight turn causes the pin to ride on the end of housing *C*, retaining locating pin *A* in a retracted position. This leaves both hands of the operator free for other duties.

If extreme accuracy is needed, a bushing *G* should be pressed into the fixture body to prevent wear. For less precise work, or for low-production jobs, this bushing can be omitted. The distance that the locating pin protrudes from the fixture body can be adjusted. To do this, lock-nut *E* is first loosened, and then knob *F* is rotated in the appropriate direction. Engagement of dowel-pin *D* in the slot will permit the housing and the shaft to be turned as one unit.

Adjustable Stock Support

When designing a work-holding fixture, it is necessary to provide support for the bottom surface of the casting or plate stock to be machined. An adjustable support is advisable so that it can always be seated snugly against the rough work surface.

An effective support of the adjustable type is illustrated in Fig. 9. This unit is built into the fixture, *A* being a section of the fixture body into which a long blind hole has been drilled and counterbored. A flanged bushing *B* is screwed and pinned in the counterbore.

Wedge extension *C* is a sliding fit within the bushing. A thumb-screw *D* is threaded in the end of the wedge extension, bearing against the end of pointed pin *E*. The pointed end of pin *E* enters a matching recess in the side of lock-pin *F*, which is free to slide in a hole drilled diametrically through extension *C*. This sub-assembly is prevented from turning by a tongue on the end of pin *F* that extends into a slot in the wall of the flanged bushing, section X-X.

A tongue on the left-hand end of the wedge extension carries a pin that engages with grooves machined in the forked end of wedge *G*. The tapered end of the wedge passes through a hole cross-drilled in plunger support *H*.

Plunger *J* is rounded at both ends, the lower end riding on the inclined surface of the wedge, and the upper end bearing against the workpiece to be supported. To prevent chips from entering the plunger and wedge assembly, cap *K* is provided.

To raise the plunger of this stock support into contact with the work, thumb-screw *D* is pushed in, but not rotated. Both the wedge and the wedge extension are thereby forced to the left. Plunger *J* rides up on the wedge surface until it contacts the work, or until it reaches the height required to accommodate the part to be supported.

The thumb-screw is then rotated to force pin *E* into the conical recess in the side of lock-pin *F*. Because the center of the conical recess is slightly below the center of pin *E*, the lock-pin is compelled to rise, jamming against the inner surface of bushing *B* and locking the assembly in position. To retract the plunger, this procedure is reversed. In the event that it becomes necessary to prevent any movement of the plunger, sockethead screw *L* can be tightened.

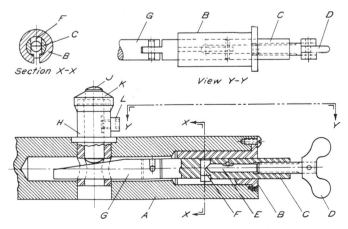

Fig. 9. Adjustable support for workpiece is built into fixture. Support plunger *J* can be raised or lowered, then locked in place, by two-directional movement of thumb-screw *D*.

Fixture Faceplate Transferred with Work from Machine to Machine

Where different machines perform a series of operations that have to be held in close relationship to each other, a system of universal two-part fixtures can be used to advantage. One of the fixtures appears in Fig. 10.

The two main parts consist of base A and faceplate B. To receive the faceplate, the vertical section of the base has a tapered hole. A short plug C on the faceplate proper has a reduced diameter corresponding to the tapered hole (3 1/2 inches per foot). Figure 11 shows the base and faceplate assembled. Keeper plate D fits over threaded stud E, so that when nut F is tightened, the faceplate is drawn firmly to the base. Spring-loaded pins G keep washer H against the nut at all times, making it easy to position the keeper plate over the stud.

Pin J, Fig. 10, retained in the vertical section of the base, engages a hole of corresponding diameter in the faceplate, to provide the proper radial location for the latter member. Instead of a single hole in the faceplate, several may be provided should indexing be required. For example, the three bosses of the workpiece fastened to the faceplate in Fig. 12 have to be drilled. The back of the faceplate contains three holes at appropriate points. The faceplate is indexed in the base for drilling each boss. Also a bushing plate is added to the top of the faceplate to guide the drill.

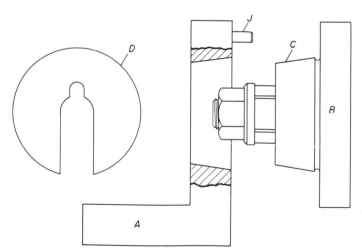

Fig. 10.   The two main parts of the fixture system are a base A for each machine in the series of operations and a faceplate B for each workpiece being processed.

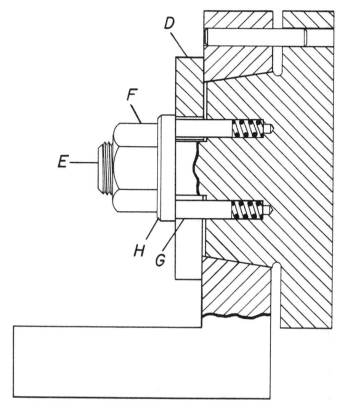

Fig. 11.   Slight rotation of nut *F* tightens or looses keeper plate *D*. Joining surfaces of base and faceplate are a fast taper, so can be readily disengaged.

A separate base is provided for, and attached to, the table of each machine involved in the required series of operations. The size and taper of the hole in the vertical section of these bases are, of course, identical. Likewise, a sufficient number of faceplates are provided, so that each workpiece being processed can remain fastened to the same faceplate without being disturbed for the entire series of operations.

For lathe work, the base design can be modified to gain rigidity, as in Fig. 13. In the example shown, the fixture is offset from the center line of the machine to bore a "clover leaf" pattern in a workpiece.

Versatile V-Block for the Diemaker Holds Unusual Shapes

Many unusual shapes and forms of piercing punches, core-plugs, pins, and similar tools or die elements are required in the manufacture and

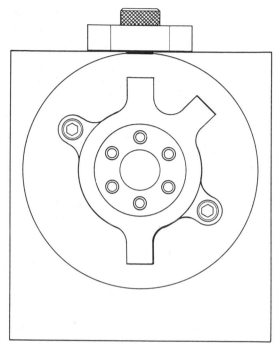

Fig. 12. The three bosses of this workpiece are aligned with the drill bushing by index-ing the faceplate. Pin in vertical section of base engages holes in back of faceplate.

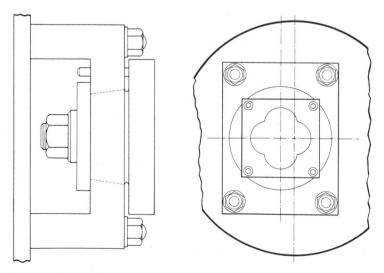

Fig. 13. This base, used on a lathe, has a U-shape for added rigidity. Faceplate is indexed four times to complete the "clover leaf."

maintenance of press tools, die-casting dies, and molds for plastics. The unusual shapes frequently present holding problems when the parts are being made. A versatile V-block specially adapted for dealing with this kind of work-holding problems is illustrated in Fig. 14. By its use parts of awkward shape can be readily gripped for light grinding or machining.

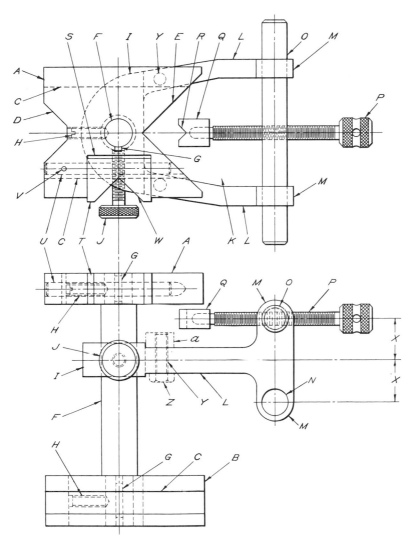

Fig. 14. Special V-block and accessories designed to facilitate handling tool and die work that is difficult to hold.

The clamping arrangements are entirely self contained in the V-block itself, so that the essential working parts cannot become mislaid when the block is not in use. Considerable adjustment is afforded by the integral clamping unit in all planes relative to the actual V-slots in the block. This enables the pressure screw to be rapidly set and presented to the workpiece in the best height, radial, or inclined relationship to suit the shape of the part. Workpieces of very short length can be accommodated that normally cannot be gripped in a standard V-block.

Referring to the diagrams of the V-block, the two hardened and ground steel plates A and B are identical except that plate B is approximately one and one-half times thicker than plate A. Extending fully along the length of the two opposed long sides of plate B are the wide slots C, by means of which the entire V-block may be clamped to the machine table.

In opposite sides of the plates A and B are the 90-degree V-slots D and E. Slots D at the left are about half the size of slots E and permit workpieces of very small diameter to be accommodated.

To insure exact duplication in the size and shape of these critical V-slots, they should be finish-ground on the inclined sides while both plates are clamped securely together. The outside edges of both plates should also be ground square and to exactly the same over-all dimensions at the same setting.

Both plates A and B are bored in the center — preferably while they are clamped together for finishing the V-slots — for mounting upon the hardened and ground cylindrical steel column F. This column is reduced in diameter a small amount at each end to provide positive shoulders for the accurate location of the two plates. These plates are secured by means of Woodruff keys G and headless screws H.

The plates are accurately mounted on the column so that the respective V-slots will be exactly in line, as shown in the upper diagram. Bored to be a free sliding fit upon the middle portion of the column is a mild steel plate I. This plate is drilled and tapped in one side wall for the brass locking screw J, which bears directly upon the side of the column. Plate I is machined with a wide slot K forming identical, parallel fork limbs L, of the same length, as shown in the upper view of Fig. 14.

The fork limbs L are made T-shaped to form the extended lugs M. These lugs are cross-drilled in exact alignment to provide two pairs of bearing holes N for the hardened and ground steel rod O, which is a free revolving fit. The rod is interchangeable in the upper and lower holes.

Rod O is drilled and tapped diametrically to receive the long pressure screw P, and is of sufficient over-all length to prevent it from passing completely out of either bearing hole N when the pressure screw is in contact with the inside wall of either limb L. Thus rod O cannot be withdrawn from its bearing holes until screw P is removed.

The center height X of the respective pairs of holes N should be about one and one-half times the thickness of the V-slotted plate A. This insures that the center of screw P, when in the horizontal plane as shown, will coincide with the top edge of plate A, when the top surface of plate I is in contact with the under side of the former member.

Quick-acting Toolmaker's Vise

A fault in many toolmakers' vises is the tendency of the sliding jaw to tilt when the work is gripped, resulting in an inaccurate and uncertain hold. The vise shown in Fig. 15 avoids this tendency, since the clamping screw A tends to force the moving jaw B down on ways provided on base C. A quick-action adjustment avoids tedious turning of a screw when setting the jaws for different sized workpieces. The elimination of the standard type of screw with a ball-ended handle makes the vise a compact unit. This design of vise has proved very convenient for mounting on sine bars and plates when laying out work on the surface plate or when machining on surface grinders or drilling machines.

The fixed jaw is integral with the vise base. The base contains a series of 3/8-inch diameter holes D, drilled completely through its width at 1/2-inch intervals. A longitudinal slot E is milled in the center of the base with a 1/2-inch end-mill. Then a recess F is milled to a depth coincident with the center line of the drilled holes. A trunnion G engages one of the semi-circular depressions formed by the milling process when the clamp screw is tightened. The trunnion nut drops out of engagement when the vise is lifted from the machine table and the clamp screw is slackened, thus allowing the sliding jaw to be quickly adjusted by hand.

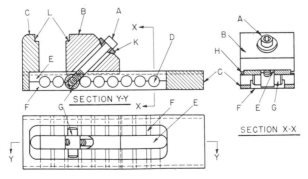

Fig. 15.   Quick-acting toolmaker's vise in which the tightening of clamping screw A tends to force movable jaw B firmly down on the base C of the vise.

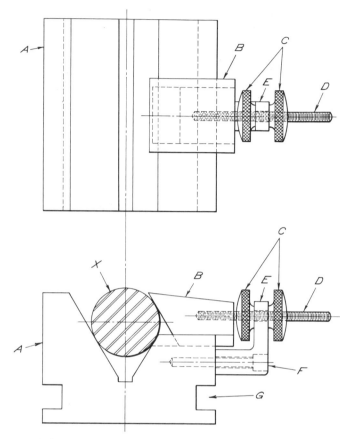

Fig. 16.   In this design, the work X is secured in the vee by the pressure of the slide B.

Lugs H, provided along both sides of the sliding jaw, guide the jaw in steps milled in the base. A spherical-seat washer K is placed under the head of the clamp screw to permit swiveling when the sliding jaw is clamping the work, and the hole in the jaw through which the screw passes is enlarged to facilitate the swiveling.

Steps L are provided at the top of the jaws for clamping thin plates. This arrangement eliminates the need for parallel strips for supporting such work during drilling operations.

V-Blocks Feature Low Head-Room Clamping

V-blocks are commonly used for supporting round work on machine tables for such operations as drilling and light milling. One objection to the use of conventional V-blocks, however, is that the straps which secure the work in the blocks require considerable head-room. If a block must be positioned close to a cutter or an arbor, the strap may obstruct the feed movement. This eventuality is eliminated in the V-block illustrated in Fig. 16.

The work X is secured in the 60-degree vee of the block A by the pressure of a slide B. The slide is dovetailed to one side of the block, which has been reduced in height. In this manner, the high point of the slide is level with the top of the opposite side of the block, and no additional head-room other than that ordinarily required by the block itself is necessary.

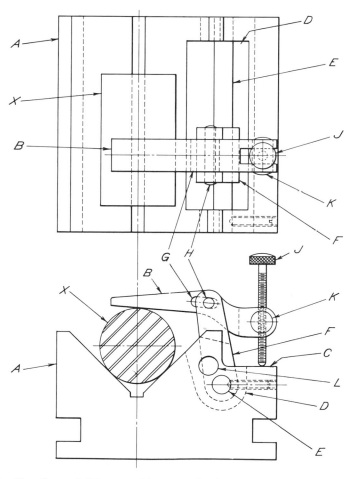

Fig. 17.   Here the work X is secured in the vee by the leverage applied by the finger B.

The left end of the slide is beveled 30 degrees so that it is parallel with the opposite wall of the vee, and the top of the slide inclines back approximately 10 degrees. Actuation of the block is by means of knurled nuts C, one for each direction. These nuts engage a threaded stud D pinned to the right end of the slide and of sufficient length to give the slide adjustment to a range of work diameters. Both nuts bear against a thrust bracket E attached to the block by cap-screws F. Slots G are provided on each side of the block for clamping it to the work-table.

Another V-block design featuring low head-room clamping is illustrated in Fig. 17. Here, the work X is secured in the 90-degree vee of the block A by the pressure of a finger B. Along the right side of the block is machined a right-angle step C. A deep curved channel D, extending for the greater part of the length of the block, is milled into the step. The left side of the channel is inclined slightly, and enters the right wall of the vee. A rod E running through both sides of the channel serves as a hinge bolt for a link F, the top of which is divided to receive the finger B. In the center of the finger is a slot G accommodating a pin H fixed across the link. As can be seen in the plan view, the link is able to be located at a point in the channel where the finger will be most effective over the work.

A lever action of the finger is instituted by turning a thumb-screw J. This screw engages a nut-block K which has a cylindrical body and which is held in a hole running transversely across a slot in the right end of the finger. When the screw is tightened down onto the top of the step C, the left end of the finger applies pressure on the work. When the screw is released, the nut-block can be swiveled slightly so that the end of the screw clears the step. The finger can then be lifted and the work unloaded. For work of relatively large diameter, the rod E can be located in a second set of holes L in the channel. By so doing, the elevation of the finger above the vee is greater. The thumb-screw is made long enough to be effective in this second position of the rod.

CHAPTER 17

Tooling Component Design—
Work-clamping

Eccentric Clamp Computations Simplified

Eccentric devices are commonly used for clamping purposes in jigs and fixtures. They are easy to produce and, if properly made, guarantee rigid clamping of the workpiece.

The tolerance on the dimension of that portion of the workpiece situated between the locating surface and the clamping surface determines the length of stroke S necessary for proper clamping with the eccentric member. The length of stroke of an eccentric clamp, in turn, depends on the eccentricity E of the clamping surface and its position angle A. In any application, however, proper clamping is obtained when the eccentric is self-locking. This can be assumed to occur if the angle of friction α between the workpiece and eccentric clamp is equal to or less than 5 degrees 43 minutes. Substituting this value in the equation $E = \dfrac{D}{2} \sin \alpha$ (where D is the diameter of the eccentric) leads to the relationship that $D \geq 20\,E$ for a self-locking eccentric clamp.

The diameter of the eccentric depends on the dimensions and the design of the jig but should be made as large as possible because of the better wear-resistance provided by a larger contact area. Position angles are limited by the operating range. It is important for the designer to have a clear conception of these factors in order to determine the proper dimensions of an eccentric clamp. Obtaining these dimensions by applying trial-and-error methods in the toolroom is both time-consuming and unnecessary. An easy way of finding the diameter and eccentricity of eccentric clamps is based on views 1, 2, and 3 of Fig. 1.

When the center of the eccentric is rotated about the center of the shaft from 0 to 180 degrees, the length of stroke increases from 0 to 2 E. In

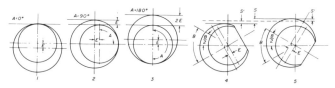

Fig. 1. Eccentric clamps should be designed for a clamping range B located between ±30 degrees of their middle position in order to obtain the greatest stroke possible for a given angle of rotation.

the middle position ($A = 90$ degrees), the length of stroke is equal to the eccentricity E. However, the length of stroke does not increase uniformly for each incremental angle during the rotation from 0 to 180 degrees. The curve of the rate of stroke increase is the curve of a harmonic motion. This curve rises very slowly up to a position angle of 60 degrees. Then the rate of increase is approximately constant up to a position angle of 120 degrees. From this angle to a 180-degree position angle the curve becomes flatter (rate of increase decreases) and levels off. Only in the range of 60 to 120 degrees does the eccentric rise uniformly for each incremental angle of rotation. Thus, for best results, the clamping range should be restricted to within -30 to $+30$ degrees of the middle position ($A = 90$ degrees). The relation between clamping range angle B and eccentricity E for a fixed length of stroke S is illustrated in views 4 and 5 which show the eccentric clamp at the two limiting position angles.

The vertical distance between the center of the eccentric and the horizontal line through the center of the shaft is equal to S', or one-half the length of stroke S, if the position angle is $\pm \dfrac{B}{2}$ of the middle position. This is obtained from the relation

$$\sin \frac{B}{2} = \frac{S'}{E}$$

and as

$$S = 2S'$$

$$\sin \frac{B}{2} = \frac{S}{2E}$$

Example:

Given: $S = 0.020$ inch, $D = 1.600$ inches, $E = 0.080$ inch

To Find: B

Solution: $\sin\dfrac{B}{2} = \dfrac{S}{2E} = \dfrac{0.020}{0.160} = 0.125.$

$$B = 14°\ 22'$$

As this clamping range angle is very small, B is chosen as 30 degrees. Then the eccentricity and diameter will become smaller.

$$E = \frac{S}{2 \times \sin\dfrac{B}{2}} = \frac{0.020}{2 \times 0.258} = 0.039 \text{ inch}$$

$$D = 20\,E = 20 \times 0.039 = 0.780 \text{ inch}$$

If the whole range of 60 degrees were to be used, there would be an eccentricity of

$$E = \frac{0.020}{2 \times 0.5} = 0.020 \text{ inch}$$

This method is a simple and easy way to obtain the most advantageous dimensions for clamping range angle, eccentricity, and length of stroke for every application of an eccentric clamp.

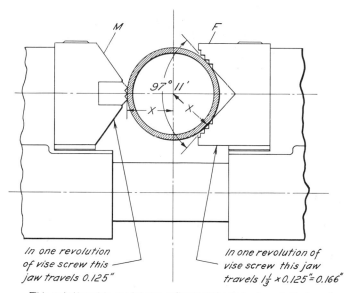

In one revolution of vise screw this jaw travels 0.125"

In one revolution of vise screw this jaw travels $1\frac{1}{3} \times 0.125" = 0.166"$

This relationship maintains alignment of stock and die head

Fig. 2.   With the lead of the jaws in the ratio of 1 1/3 to 1, jaw F has an included angle of 97 degrees 11 minutes.

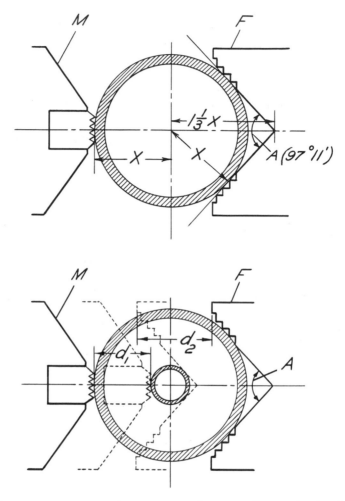

Fig. 3.   To center the work, jaw *F* must travel a greater distance than jaw *M*.

Finding Vee Angle of Female Jaw for Self-Centering Vise

A new self-centering vise on a pipe-threading machine features a pair of mating vise jaws, as can be seen in Fig. 2. Both jaws are engaged to a combination left- and right-hand screw, and can grip pipe and tubing ranging in diameter from 1/4 inch to 2 1/2 inches. They replace a pair of symmetrical jaws that had a deep vee on one side and a shallow vee on the other — a design that was necessary for the same range of work, but which often required reversing the jaws on a change-over from one size of pipe to another.

There is a peculiarity of the new design in centering the work with the pipe die. Although both jaws are engaged to the same screw, the pitch of the screw section for the female jaw F must be coarser than that for the male jaw M. This point is illustrated in Fig. 3. If, for example, a 1/2-inch pipe is to replace a 2-inch pipe in the vise, jaw M moves in the distance d_1, or 0.768 inch (exactly one-half the difference between the outside diameters); but jaw F moves in the greater distance d_2.

From an engineering and machining standpoint, it was found practical to specify 8 threads per inch for the male section of the screw, and 6 threads per inch for the female section. Thus, the lead of the female jaw in relation to the male jaw is in the ratio of 1 1/3 to 1. To determine the included, or vee angle A of the female jaw so that for the given combination of 8 and 6 threads per inch any diameter of work can be centered with the pipe die, this equation was used:

$$\sin 1/2\ A = \frac{X}{1\ 1/3\ X}$$

Solving for A,

$$\sin 1/2\ A = 0.75$$

$$1/2\ A = 48 \text{ degrees } 35 \text{ minutes } 30 \text{ seconds}$$

$$A = 97 \text{ degrees } 11 \text{ minutes}$$

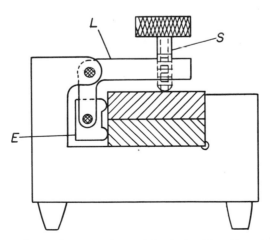

Fig. 4. Rectangular sections are clamped firmly in this milling fixture by two-way pressure exerted by screw S and lever L.

Handy Fixture Clamps Effective with Simple Shapes

Fast, positive action is the keynote of two clamping devices, with a fixture (Fig. 4) holding two rectangular sections at once, and a jig (Fig. 5) arranged to permit drilling both legs of an angle section accurately.

The milling fixture clamp (Fig. 4) holds two parts at once, with pressure to the right distributed by equalizer shoe E, and downward from the simple knurled set-screw S on the lever L.

Angles are held for drilling operations on either leg of right-angle stock, using the box jig in Fig. 5. When wrench A is tightened on finger strap B, pressure forces the work to the left and upward simultaneously. The work is held securely, and if end-stops are provided, work cannot be wrongly positioned.

Double-Action Clamping Device

The clamping device shown in Fig. 6 is designed with a double-action holding arrangement by means of which pressure can be applied simultaneously on a cylindrical part and on a collar mounted on the part. As illustrated, a shoulder on shaft A is forced against a locating surface on a fixture and collar B is clamped against a second locating surface.

When the knurled knob C is rotated, clamping member D is forced against the end of the shaft, as seen in the top view. Member D then pivots on its semicircular clamping surface, and transmits force through shaft E to the outer member F. The latter member is a fork which encloses member D. Both members are cross-pinned together by the shaft E.

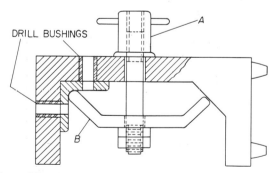

Fig. 5. Both legs of "angle iron" can be drilled in this economical, versatile box jig. Construction is so simple and inexpensive that they can be made in a variety of sizes and hole patterns for short, repetitive lots of work.

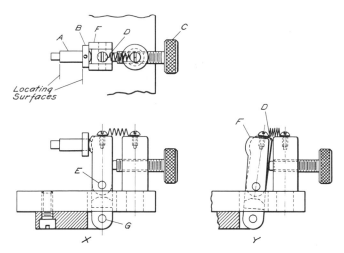

Fig. 6.   Device providing a double clamping action through the use of coupled links.

Pivoting on shaft *G*, the outer clamping member straddles the shaft and exerts pressure against the face of the shaft collar at two diametrically opposed points. A tension spring seen at the top of the device returns the clamp assembly to its original setting when the knurled knob is backed off. View Y shows the device in the unclamped setting and view X in the clamped setting. The outline of the inner clamping member is more clearly discernible in view Y.

Multiple-Ball Clamping Compensates for Part Variations

An effective method of clamping more than one part of a kind for machining, irrespective of small tolerance variations, makes use of multiple balls. The principle here is that pressure applied to the balls will cause them to realign themselves in the face of unequal resistance. A clamping setup designed along this line is illustrated in Fig. 7.

Workpieces *A* (in this case, cylindrical) are held in position over plungers *B* by V-grooves in the face of jaw *C*. The flanged rear ends of the plungers extend into a recess in housing *D* and are in contact with three steel balls *E* of bearing quality.

To clamp the work, pressure is applied to balls *E* through two additional balls *F* — of like quality — and pressure-plate *G*. Any size variation that may exist between the parts being clamped will be compensated for by a slight shift in the relative positions of the five balls, thus equalizing the pressure transmitted to the individual workpieces. Springs *H* cause the plungers to withdraw upon release of the clamping force.

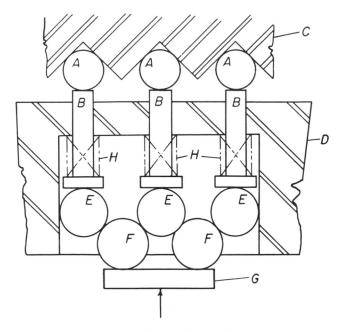

CLAMPING FORCE

Fig. 7. A multiple-ball principle is used in this unit for clamping more than one work-piece. Tolerance variations between the parts are compensated for by a slight shifting of the balls when clamping pressure is applied.

Vise Jaws Self-adjusting for Multiple Parts

Occasionally, the milling of three or more parts at one time is advantageous. Self-adjusting vise jaws for simultaneously holding three cylindrical workpieces of slightly different diameters are illustrated in Fig. 8. These special jaws are easily attached to a machinist's vise by cap-screws.

Fixed jaw A, shaped as shown, has three vertical, 90-degree V-grooves. Movable jaw B is equipped with three horizontal plungers C situated opposite the V-grooves, and on closing the vise each workpiece D is clamped in one of the V-grooves by a plunger. Plungers C move in and out to take care of variation in part diameter. This small movement is compensated automatically by a lateral adjustment of the angle-faced plungers E. Pins F serve to transmit motion from plunger to plunger. These pins are mounted in clearance holes to permit the movement of the plungers. Plate G is removable to permit lubrication of the moving parts.

When the vise jaws are assembled for use, pins *F*, should be in the center of their clearance holes. This is accomplished by adjusting the two large slotted set-screws *H* and moving plungers *C* in or out as necessary. Retaining screws *J* should be replaced temporarily by longer screws which can be used to clamp plungers *C* in place after they are properly adjusted. The clamping ends of the three plungers should then be ground true and parallel to the face of the jaws in one setup. After inserting retaining screws *J*, the jaws are attached to a machinist's vise *K* for milling workpieces with cutter *L*. Although these vise jaws were designed to hold three parts, redesign for additional plungers and pins will increase the part-holding capacity.

Light-Duty Pivoting Clamp

To locate a workpiece in a fixture for an assembly operation or a machining operation in which pressures are light, the quick-acting clamp, shown in Fig. 9, can be used.

Clevis *A* extends from the body of fixture *B*. Clamp *C* pivots on dowel-pin *D* press-fitted in the clevis. Compression spring *E* is contained between two ball studs *F*, one being in the fixture and the other secured as illustrated in short leg *G*.

The view at the left shows the device in clamping position, with long leg *H* applying pressure on workpiece *J*. To unclamp the workpiece, as in the view at the right, handle *K* is pushed forward to stop-pin *L*. The base of the fixture beneath the stop-pin is cut off at an angle to provide finger clearance for pulling the handle forward.

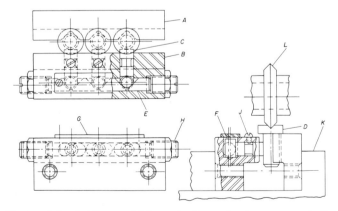

Fig. 8. Designed for holding multiple parts, these vise jaws automatically adjust to slight size variation in the workpieces.

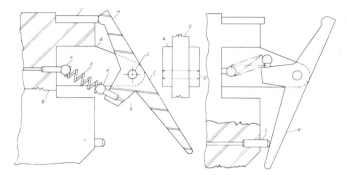

Fig. 9. In clamping position, spring *E* rotates long leg *H* toward workpiece *J*. When the clamp is pivoted, the spring rotates the long leg away from the workpiece.

Brake Band Locks Positioning Fixture

Welding, brazing, assembly, and other similar type fixtures very often have to be adjusted to numerous positions while performing various operations. Complicated plunger type indexing mechanisms are usually slow acting as well as unwieldly and difficult to operate. Illustrated in Fig. 10, however, is a simple and inexpensive braking arrangement that facilitites positioning, thus reducing operator fatigue while helping to increase production.

To make use of this braking arrangement, the fixture housing is bored to receive ball, roller, or needle bearings. These can be of either the open or sealed type, depending on the application of the fixture. The

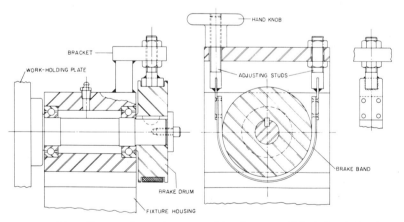

Fig. 10. Locking of rotary fixture in any position is accomplished simply by the use of a brake drum and adjustable brake band. The brake drum is keyed to the shaft carrying the work-holding plate.

end of the main shaft opposite the work-holding plate is turned and a keyway is milled to accept a single brake drum as shown.

A bracket welded to the top of the housing has two clearance holes drilled through it to accommodate brake-band adjusting studs. The bottom ends of the two studs are slotted and have mild-steel strips welded to them which are, in turn, riveted to the brake band. The brake band itself is of simple construction. It is made by riveting a length of standard brake lining to a backing sheet of either brass or steel.

Two jam nuts on the right-hand stud not only lock it to the bracket, but also permit adjustment of the brake band. In operation, slight tightening of the hand knob on the left-hand stud will clamp the work-holding plate securely in any position desired.

Finger Clamp for Gripping Non-parallel Surfaces

One shortcoming of the ordinary toolmaker's finger clamp is its inability to grip non-parallel surfaces effectively. Because only one of its two jaws has a proper bearing, it is necessary to insert shims or other blocking material in the open area between the other jaw and the work. A more practical design of finger clamp for work of this nature is shown in Fig. 11. In the illustration, one side of a wedge-shaped part *A* bears against the upper jaw *B* of the clamp, and the other side bears against a

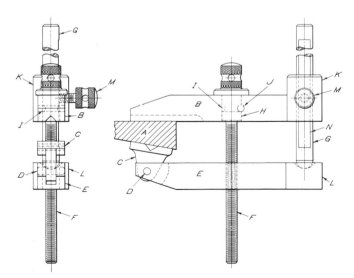

Fig. 11. A finger clamp incorporating a swivel-block provides an effective grip for non-parallel surfaces.

block *C*. This block swivels on a pin *D* held in a slot milled into the front of the lower jaw *E*.

Instead of the usual two-screw arrangement of the ordinary finger clamp, this clamp has a single screw *F* running through the centers of the jaws, and a pillar *G* aligning the back ends of the jaws. The greater part of the length of the screw is threaded and engages a tapped hole in the lower jaw. A shoulder *H* fits a reamed hole in the upper jaw. To prevent the screw from rising in the upper jaw when the clamp is opened, the shoulder is provided with an annular groove *I*, which is intersected by a pin *J* having a press fit in a hole through the jaw.

The back end of each jaw is formed into an enlarged square. Pillar *G* runs completely through a hole in the square end *K* of the upper jaw and has a ball-like seat (to maintain both jaws in the same vertical plane) in the square end *L* of the lower jaw. For the screw to function properly, it is necessary to keep the jaws parallel. This is accomplished by adjusting the position of the upper jaw on the pillar by means of a thumb-screw *M* which is tightened against a milled flat *N*.

Although primarily designed for gripping non-parallel surfaces, this finger clamp has other practical uses, some of which are illustrated in Fig. 12. The jaws are shown gripping a part having parallel surfaces in (*a*) and gripping a U-shaped member and a flat member in (*b*). Also, because the block has a crosswise vee and the upper jaw a lengthwise vee,

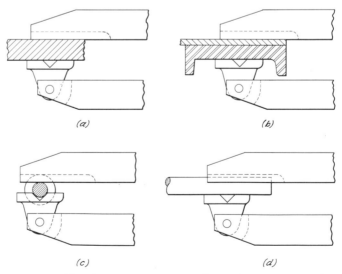

(*a*)   (*b*)

(*c*)   (*d*)

Fig. 12. Several additional practical uses of the finger clamp are shown in this illustration.

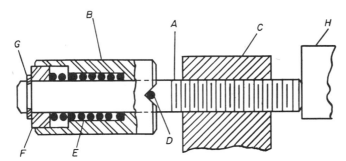

Fig. 13.   Torque screw *A* and barrel *B* rotate as a unit until pressure against the work *H*
stalls the torque screw, thrusting the barrel back.

round stock can be held at right angles to the jaws as in (*c*) or parallel
to the jaws as in (*d*).

Clamping Plunger Applies Controlled Pressure Outside of Operator's Influence

Setups in which work must be secured lightly with a preset amount of
pressure can include one or several of the clamping plungers illustrated
in Fig. 13. The two main elements of the device are torque screw *A* and
barrel *B*. An important advantage of this device is that the pressure
applied is not controlled by the operator; therefore it is always the same
from piece to piece, and from operator to operator.

Fixed to the jig, fixture, or machine table is a block *C* having a tapped
hole which engages the torque screw. Pin *D* is a press fit in a hole cross-
drilled in the torque screw a short distance behind its threaded section.
The barrel encloses the torque screw, and has a V-groove milled across
its face. Inside, the barrel is bored to receive spring *E* and also counterbored to receive ring *F*, which is retained in position near the end of
the torque screw by clip *G*.

To clamp the work *H*, the barrel is rotated. Since the spring keeps the
V-groove over the pin, the torque screw is forced to rotate as a unit
with the barrel, and is advanced through the block. Then, when continued movement of the torque screw against the work builds up sufficient pressure, the barrel is thrust back, and the pin rides on the face of
the barrel. Thus, the torque screw remains stationary as the barrel is
rotated further.

If, after the clamping plunger is constructed, it is decided to increase
or decrease the pressure on the work, another spring can be used.

Index

BILL BECK.

901 682 — 4 875

125

200

1 = B ——

125 =

9.00

1 (20.00) 6½ X

10. —

12 X 30

1 —

125. — Chuck Mounted.

10. Live Center.

4 Colts Each —

625